LIPID
BIOCHEMISTRY

LIPID BIOCHEMISTRY

For Medical Sciences

ASHOUR SALEH ELJAMIL

 iUniverse®

LIPID BIOCHEMISTRY
FOR MEDICAL SCIENCES

iUniverse books may be ordered through booksellers or by contacting:

iUniverse
1663 Liberty Drive
Bloomington, IN 47403
www.iuniverse.com
1-800-Authors (1-800-288-4677)

ISBN: 978-1-4917-4223-5 (sc)
ISBN: 978-1-4917-4222-8 (e)

Library of Congress Control Number: 2014917826

Print information available on the last page.

iUniverse rev. date: 03/30/2015

With love and respect to my wife, Miriam; my son, Mohamed; and my daughters, Esra, Abbrar, and Ala; and to the spirit of my dearest daughter, Aya.

Ashour Saleh Eljamil, 2015

FOREWORD

Current knowledge and understanding of lipid biochemistry has evolved rapidly in recent decades. This has happened because prior to recent decades, lipids were not a popular area of study except by hardy chemists who did not mind working with large volumes of flammable and often toxic organic solvents (such as methanol, chloroform, and ether) and for whom the methods of fractionation commonly used with water-soluble compounds such as proteins and carbohydrates were not suitable. However, the development of a wide variety of chromatographic methods in the second half of the last century opened up the subject for the study of the wide diversity of lipids of human and animal origin. This has made it possible to move from the study of adipose tissue, which consists mostly of triacylglycerols (which are storage lipids) to that of other tissues in which the lipids are mostly polar and structural in function (such as phospholipids, sphingolipids, cerebrosides, and gangliosides). The combination of gas-liquid chromatography and mass spectrometry played a key role in this amazing growth.

It is not common knowledge that the non-water components (i.e. about 35%) of the theoretical 70 kg male human body consist of approximately equal amounts of lipid (about 13%, or 9 kg) and protein (about 15%, or 11 kg) and the rest of skeletal and other minerals, carbohydrates, and nucleic acids. In severe obesity, the lipid content can account for up to about 70% (or 57 kg) of body weight, mostly in subcutaneous adipose tissue. The ease with which extra groceries are converted into storage fat contrasts sharply with the difficulty that many obese individuals have in getting rid of their excess store of fat.

Readers of this book are likely to be more interested in its factual contents than in the history of their acquisition. This is natural and understandable but may lead readers to form the impression that these are the products of recent decades only. This would be unfair to the many inventive pioneers who prepared the grounds for much of our recent knowledge. Among these were William Prout, the English physician, and Claude Bernard, the illustrious French physiologist. In 1827, the former classified the organic components of human diet into saccharine, oily, and albuminous, or, in more modern terminology, carbohydrate, lipid, and protein. Between 1846 and 1856, the latter demonstrated the lipase activity of the rabbit pancreas, showing that pancreatic juice breaks down fat into glycerol and fatty acids and that pancreatic juice and bile are essential for the digestion and absorption of fat in the duodenum. Unsophisticated their methods may have been, but very significant and enduring were their conclusions.

The proteins are encoded by the bulk of the genes in DNA and provide the machinery for the production of body lipids from the groceries we consume. This machinery consists of enzymes, carriers, hormones, agents for gene regulation, and such. Elucidation of what it is that the different structural lipids specifically do in the membrane systems of the body cells, especially those of the central nervous system, is a gigantic task that awaits future research. Although they are not direct gene products, the lipids deserve our attention and admiration.

This book is characterized by its comprehensive survey of the subject, the brevity of its descriptions, and its primary interest in making the subject relevant to readers interested in the molecular aspects of the working of their bodies. It is a valuable guide to the second-largest component of the non-obese human body and to some of the medical problems that may afflict the body in its handling of this fascinating component.

Francis Vella
M.D., M.A.(Oxon), Ph.D., hon D.Sc.
Retired Professor of Biochemistry, University of Saskatchewan, Canada

PREFACE

This book introduces lipid biochemistry to undergraduate students in the medical sciences without overloading the description with unnecessary details. Understanding biochemistry requires familiarity with the structure of biomolecules, as this provides a visual presentation of their chemistry and, frequently, of their function. For this reason, the first chapter places special emphasis on the individual chemical structure of the various lipid classes. The digestion and absorption of dietary lipids is briefly discussed in the second chapter. Aspects of their metabolism is presented in separate chapters on fatty acid synthesis, fatty acid oxidation, acylglycerols and sphingolipids, glycolipids, cholesterol, plasma lipoproteins, steroid hormones, and fat-soluble vitamins.

The synthesis of saturated, monounsaturated, and polyunsaturated fatty acids and eicosanoids are described, but emphasis is given to the saturated type. The chapter dealing with the oxidation of fatty acids starts with the mobilization of stored fat, continues with the pathways of fatty-acid oxidation, and concludes with the diseases resulting from impaired oxidation of fatty acids.

Chapter 5 simply describes the biosynthesis and degradation of phospholipids (acylglycerols and sphingophospholipids).

Glycolipid metabolism is also described, in chapter 6, with special emphasis given to lipid-storage diseases.

Chapter 7 describes cholesterol metabolism, with particular attention to its synthesis and regulation, and describes the degradation of cholesterol to bile salts.

Chapter 8 gives attention to the metabolism of lipoproteins, including their formation and breakdown. This chapter also discusses diseases of lipoprotein metabolism.

The functions, biosynthesis, transport, mechanism of action, and excretion of steroid hormones is the topic of chapter 9.

Finally, chapter 10 presents the biochemical actions, deficiency, and toxicity of fat-soluble vitamins.

ACKNOWLEDGEMENTS

I am grateful to Professor Frank Vella (Biochemistry Department, College of Medicine, University of Saskatchewan, Saskatoon, Saskatchewan, Canada) for his remarkable suggestions, remarks, and input throughout this book. I am also very grateful to Professor Khaled M. Elnageh (Biochemistry Department, Faculty of Medicine, Tripoli University, Tripoli, Libya) and to Professor Fared Alasmer (Biochemistry Department, Faculty of Medicine, Ain Shams University, Cairo, Egypt) for their suggestions and excellent ideas to make the present text informative and readable.

CONTENTS

Chapter 1: Lipid Chemistry

Introduction

Lipids are a heterogeneous group of organic molecules that are water-insoluble (hydrophobic) and soluble in nonpolar solvents such as benzene, chloroform, and ether. Biological lipids are a subgroup of these molecules that occur naturally and include fatty acids (FA), acylglycerols, phospholipids, sphingolipids, sterols, eicosanoids, terpenes, prenol, fat-soluble vitamins, and waxes. Lipids are divided into eight categories: fatty acyls, glycerolipids, glycerophospholipids, sphingolipids, saccharolipids, polyketides, prenol lipids, and sterol lipids. These play important roles in energy metabolism and in the function and structure of biological membranes. This chapter focuses on the classification, structure, and functions of biological lipids. The structure and function of fat-soluble vitamins will be discussed in chapter 10.

Functions of Lipids

In the body, lipids provide:
- A major source of energy (triacylglycerols[TAG]).
- Thermal insulation in the subcutaneous tissues and around certain organs (TAG).
- Part of the structure of cell membrane (phospholipids, cholesterol).
- Regulatory or coenzyme activity (fat-soluble vitamins, eicosanoids, steroid hormones).
- Control of homeostasis (prostaglandins and steroid hormones).
- Constituents of nervous tissue structure (sphingolipids).
- Regulation of growth (essential fatty acids).
- Solution of other lipids during digestion (bile acids).
- Aid to signal transduction processes such as immunoglobulin-E signalling, T-cell antigen receptor signalling, B-cell antigen receptor signalling, epidermal growth factor (EGF) receptor signalling, and insulin receptor signalling (lipid rafts).

Classification of Lipids

In 2005, The International Lipid Classification and Nomenclature Committee (ILCNC) established a comprehensive classification system for lipids, which are small hydrophobic or amphipathic molecules that may originate entirely or in part from carbanion-based condensations of thioesters (fatty acyls, glycerolipids, glycerophospholipids, sphingolipids, saccharolipids, and polyketides) or by carbocation-based condensations of isoprene units (prenol lipids and sterol lipids). Each of the eight categories that ILCNC established, which are listed below, is further divided into subclasses. The chemistry of these biologically important lipids will be presented in each section below.

1) FATTY ACYL (FA)

The *fatty acyl* group is divided into classes of fatty acids and conjugates, including octadecanoids, eicosanoids, fatty alcohol, fatty aldehydes, and fatty esters. Fatty acid structures are important building blocks of complex lipids, including biological lipids. The carbon chain, normally four to twenty-four carbon atoms long and either saturated or unsaturated, may be attached to functional groups containing oxygen, halogens, nitrogen, or sulphur. The structures of some selected fatty acids are shown in figure 1.1.

Nomenclature

The systematic name for a fatty acid is derived from the name of its hydrocarbon tail but with the substitution of –oic for the final –e. Thus, saturated acids end in *anoic* (e.g. oct*anoic* acid) and unsaturated acids with double bonds end in *enoic* (e.g. one C18 fatty acid with two double bonds is octadecadi*enoic* acid). The symbol of 20:0 denotes a C20 fatty acid with no double bonds, whereas 20:3 indicates that there are three double bonds. Fatty-acid carbon atoms are numbered starting from the carboxyl carbon as carbon 1. Carbon atoms 2 and 3 are often referred to as α and ß, respectively, and the end methyl carbon is the ω-carbon or n-carbon atom. The R/S designations (as opposed to α/β or D/L) can be used to define stereochemistries, each chiral center in a molecule is assigned a prefix (R or S), according to whether its configuration is right- or left-handed. The symbol R comes from the Latin *rectus* for right, and S from the Latin *sinister* for left.

The position of the double bond is represented by the symbol Δ and a superscript number. For example, *cis*-Δ^7 means that there is a *cis* double bond between carbon atoms 7 and 8. The position of a double bond can also be denoted by its position from the methyl terminal, starting from the ω-carbon atom. For example, ω-6 indicates a double bond on the sixth carbon. The number and the positions of the double bonds are shown in figure 1.2.

In animals, new double bonds are introduced only between the existing double bond (e.g. ω-9, ω-6, or ω-3) and the carboxyl carbon, creating three series of fatty acids, known as the ω-9, ω-6, and ω-3 families, respectively. Fatty acids vary in chain length and degree of unsaturation. In biological systems, they usually contain an even number of carbon atoms, and in animals, the hydrocarbon chain is almost invariably unbranched. This type of double bond can be indicated by the *E/Z* system (as opposed to *cis/trans*). If the two groups with the higher priorities are on opposite sides of the double bond, this is referred to as E isomer, but if the two groups with the higher priorities are on the same side on the double bond, that is described as Z isomer. The E from German entgegen; opposite, whearse the Z from German zusammen; together.

Saturated fatty acids

These lipids contain a straight hydrocarbon chain (a hydrophobic tail) with a terminal carboxylate group (a hydrophilic head), with the structural formula $CH_3 - CH_2 - CH_2 - CH_2 - CH_2 - (CH_2)_n - COOH$. Palmitate (16:0) and stearate are the most abundant saturated fatty acids in human tissues. Table 1.1 shows these and other saturated fatty acids present in mammalian tissues.

Eicosanoids (C20), another class of the fatty acyl group, are unbranched fatty acids with multiple functional groups derived from arachidonic acid. This class includes prostaglandins, thromboxanes, and leukotrienes.

The *fatty alcohol* and *fatty aldehyde* classes of the acyl category are characterized by terminal hydroxy and oxo groups, respectively. *Fatty esters* and *fatty amides* are also classes of fatty acyls. Fatty esters include some biochemically important intermediates such as wax esters, fatty acyl-CoA derivatives, fatty acyl carrier protein (ACP) derivatives, and fatty acyl carnitines. The fatty amides class includes primary amides and acyl amides.

Branched-chain fatty acids

These are straight-chain fatty acids such as phytanate with one or more methyl substituents (*see* figure 4.8)

Unsaturated fatty acids

These fatty acids contain one or more double bonds (*see* table 1.2).

Monounsaturated fatty acids

These fatty acids contain one double bond. Oleic acid, the most common unsaturated fatty acid in mammals, is an example.

Polyunsaturated fatty acids (PUFAs)

These contain more than one double bond. For example, arachidonic acid contains four double bonds. In naturally occurring fatty acids, the double bonds are *cis* and are separated by one methylene group (e.g. $CH=CH-CH_2-CH=$).

Figure 1.1

Structures of some fatty acids: (a) palmitic acid, (b) oleic acid, and (c) linoleic acid.

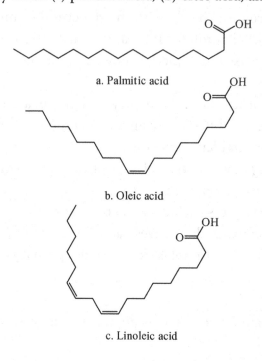

a. Palmitic acid

b. Oleic acid

c. Linoleic acid

Table 1.1

The number of carbon atoms and the structural formula of some important saturated fatty acids.

Number of carbon atoms	Common name	Formula
2	Acetic	CH_3-COOH
3	Propionic	CH_3-CH_2-COOH
4	Butyric	CH_3-$(CH_2)_2$-COOH
6	Caproic	CH_3-$(CH_2)_4$-COOH
8	Caprylic	CH_3-$(CH_2)_6$-COOH
10	Capric	CH_3-$(CH_2)_8$-COOH
12	Lauric	CH_3-$(CH_2)_{10}$-COOH
14	Myristic	CH_3-$(CH_2)_{12}$-COOH
16	Palmitic	CH_3-$(CH_2)_{14}$-COOH
18	Stearic	CH_3-$(CH_2)_{16}$-COOH
20	Arachidic	CH_3-$(CH_2)_{18}$-COOH
24	Lignoceric	CH_3-$(CH_2)_{22}$-COOH

Essential fatty acids (EFAs)

In 1930, Burr and Burr reported that acute deficiency symptoms could be produced in rats by feeding them fat-free diets and that these symptoms would disappear after the addition only of arachidonic acid and linoleic acid, the latter of which, along with α linolenic acid (see table 1.2), cannot be synthesized in the body so must be supplied in the diet. Hence, they are called *essential* fatty acids. The average daily requirement for an adult is 2 to 10 g of linoleic acid. EFAs are important structural and functional components of cell membranes and serve as precursors of eicoasanoids.

....

Figure 1.2

Linoleic acid; ω_6, C18: 2; or Δ^9, ^{12}18: 2, or n–9, 18: 2

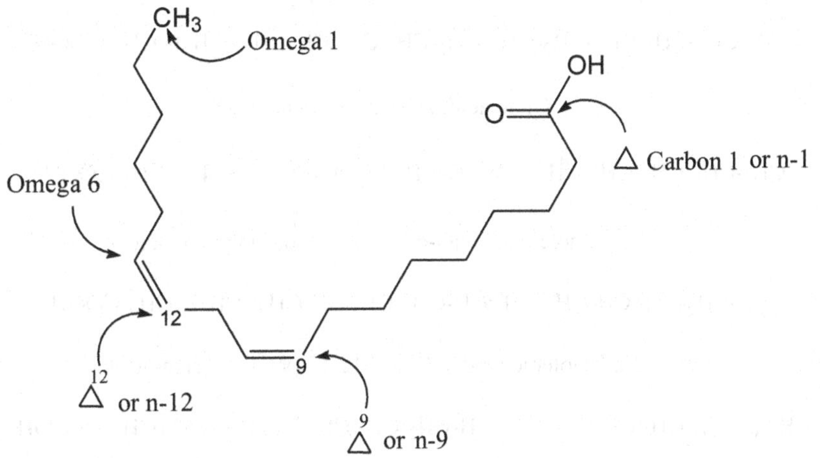

Arachidonic acid becomes essential if its precursor, linoleic acid, is missing in the diet. Animals can introduce double bonds at Δ^4, Δ^5, Δ^6, and Δ^9 positions (counting from the carboxyl terminal) but not beyond the Δ^9 position. However, plants can introduce new double bonds beyond the Δ^9 position and so are able to synthesize the essential fatty acids.

EFAs also serve as precursors of eicosanoids. Further, they prevent deposition of cholesterol beneath the endothelium of blood vessels by increasing its solubility, so they protect against the formation of atherosclerosis.

Eicosanoids

Eicosanoids are derivatives of eicosa- (C20) polyunsaturated fatty acids and comprise prostanoids and leukotrienes (LTs). Prostanoids include prostaglandins (PGs), prostacyclins (PGI), and thromboxanes (TXs). Figure 1.3 shows the structures of some eicosanoids. These compounds are synthesized by all mammalian tissue, where they act as local hormones, and they have a wide range of important physiologic and pharmacologic activities. PGs are formed in vivo by cyclization of the centre of the carbon chain of C20 polyunsaturated fatty acids (e.g. arachidonic acid and eicosapentaenoic acid) to form a cyclopentane ring structure. Thromboxanes are oxygenated eicosanoids that differ from prostaglandins in their oxane-ring structure.

Eicosanoids formed from three different unsaturated fatty acids, characterized by the number of double bonds in the side chain, are called PGE_1, PGE_2, and PGE_3. Variations in the substituent groups attached to the rings give rise to different types in each series of prostanoid. Leukotrienes, first described in leukocytes, are formed by the lipoxygenase pathway and characterized by the presence of three conjugated double bonds.

Table 1.2

Structure of some unsaturated fatty acids of physiologic importance.

$$CH_3\text{-}CH_2\text{-}CH_2\text{-}CH_2\text{-}CH_2\text{-}CH_2\text{-}CH{=}CH\text{-}CH_2\text{-}(CH_2)_5\text{-}CH_2\text{-}COOH$$

Palmitoleic, cis-9-hexadecenoic acid

$$CH_3\text{-}CH_2\text{-}CH_2\text{-}CH_2\text{-}CH_2\text{-}CH_2\text{-}CH_2\text{-}CH_2\text{-}CH{=}CH\text{-}CH_2\text{-}(CH_2)_5\text{-}CH_2\text{-}COOH$$

Oleic acid, cis-9-octadecenoic acid

$$CH_3\text{-}CH_2\text{-}CH_2\text{-}CH_2\text{-}CH_2\text{-}CH{=}CH\text{-}CH_2\text{-}CH{=}CH\text{-}CH_2\text{-}(CH_2)_5\text{-}CH_2\text{-}COOH$$

Linoleic acid, all-cis-9.12-octadecadienoic acid

$$CH_3\text{-}CH_2\text{-}CH{=}CH\text{-}CH_2\text{-}CH{=}CH\text{-}CH_2\text{-}CH{=}CH\text{-}CH_2\text{-}(CH_2)_5\text{-}CH_2\text{-}COOH$$

a-Linolenic acid, all-cis-9.12.15-octadecatrienoic acid

$$CH_3\text{-}(CH_2)_4\text{-}CH{=}CH\text{-}CH_2\text{-}CH{=}CH\text{-}CH_2\text{-}CH{=}CH\text{-}CH_2\text{-}CH{=}CH\text{-}(CH_2)_3\text{-}COOH$$

Arachidonic acid, all-cis-5.8.11.14-eicosatetraenoic acid

$$CH_3\text{-}(CH_2\text{-}CH{=}CH\text{-})_5\text{-}CH_2\text{-}CH_2\text{-}CH_2\text{-}COOH$$

Eicosapentaenoic acid, all-cis-5.8.11.14.17-eicosapentaenoic acid

Figure 1.3

Structure of some eicosanoids.

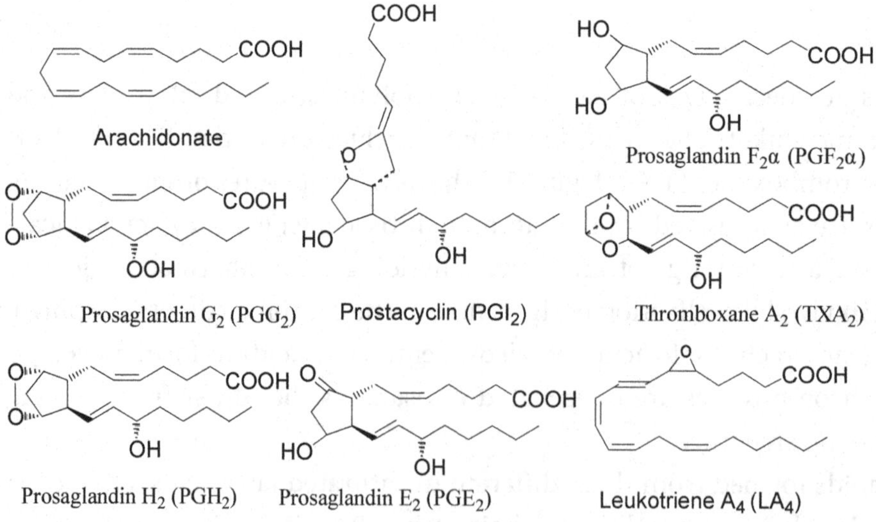

Arachidonate

Prosaglandin F$_2\alpha$ (PGF$_2\alpha$)

Prosaglandin G$_2$ (PGG$_2$)

Prostacyclin (PGI$_2$)

Thromboxane A$_2$ (TXA$_2$)

Prosaglandin H$_2$ (PGH$_2$)

Prosaglandin E$_2$ (PGE$_2$)

Leukotriene A$_4$ (LA$_4$)

2) GLYCEROLIPIDS (GL)

Glycerolipids are divided into three classes, which are further divided into subclasses. The most common glycerolipids are the mono-, di-, and tri-substituted glycerols; they are fatty acid esters of glycerol (acylglycerols).

Triacylglycerol (TAG), monoacylglycerol (MAG), and diacylglycerol (DAG) are usually referred to as *neutral lipids*. They are the storage forms of fatty acids. Naturally occurring fats are nearly all acylglycerols containing more than a kind of fatty acid (e.g. distearopalmitin). Examples of acylglycerols are shown in figure 1.4. Glycerolglycans are characterized by the presence of one or more sugar residue linked to the glycerol moiety through glycosidic linkage.

3) GLYCEROPHOSPHOLIPIDS (GP)

Glycerophospholipids (GPs) are polar compounds that consist of an alcohol attached by a phosphodiester bridge to diacylglycerol. They are abundant in nature and are essential components of the lipid bilayer of cell membranes. Based on the nature of the polar-head group at the sn-3 position of the glycerol backbone phospholipids are subdivided into distinct classes: phosphatidic acid, phosphatidylglycerol, phosphatidylcholine (lecithin), phosphatidylethanolamine (cephalin), phosphatidylinositol, phosphatidylserine, lysophospholipids, and plasmalogens. Examples of GP structures are shown in figures 1.5 and 1.6.

GPs are formed from phosphatidic acid (PA) and an alcohol. The phosphate group on PA can be esterified to another molecule of choline, ethanolamine, glycerol, inositol, or serine to produce phosphatidylcholine (PC), phosphatidylethanolamine (PE), phosphatidylglycerol, phosphatidylinositol (PI), or phosphatidylserine (PS). Removal of the fatty acid at either carbon 1 or 2 of a phosphoglyceride forms a lysophosphoglyceride such as lysolecithin.

Cardiolipin

Cardiolipin so named because it was first isolated from an animal heart and also termed diphosphatidylglycerol (see figure 1.6), is located primarily in the inner mitochondrial membrane. Cardiolipin is essential for the function of several enzymes of oxidative phosphorylation and thus is required for energy production, particularly in the heart muscle. In the inner mitochondrial membrane, it serves as Ca^{2+} binding site through which Ca^{2+} triggers increased membrane permeability. It is also involved in maintaining mitochondrial membranes' potential and their osmotic stability and importing protein into mitochondria and required for cell survival. Cardiolipin is formed when two molecules of phosphatidic acid are esterified through their phosphate groups to an additional molecule of glycerol (figure 1.6).

Plasmalogens

When a fatty acid is attached to a glycerol by an ether rather than by an ester linkage at carbon 1, a plasmalogen such as phosphatidal ethanolamine, which is similar in structure to phosphatidylethanolamine (figure 1.7), is produced.

Figure 1.4

Examples of triacylglycerols (TAG)

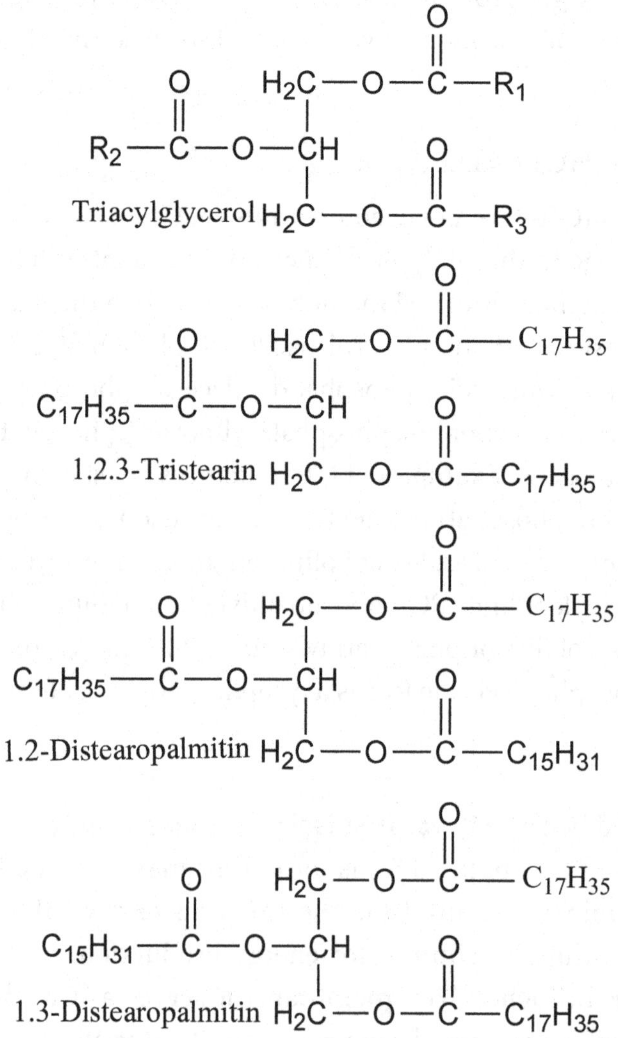

Major functions of glycerophospholipids (GPs)

- GPs are essential components of all cell membranes.
- Dipalmitoyl lecithin is the major lipid component of the lung surfactant in the extracellular fluid lining the alveoli. This surfactant decreases the surface tension of this fluid layer, thereby preventing alveolar collapse. An absence or deficiency of this surfactant in the lungs of premature infants causes respiratory distress syndrome.

- Phosphatidylinositol 4, 5-bisphosphate can be cleaved into 1, 2-diacylglycerol and inositol trisphosphate, which are second messengers.
- The platelet-activating factor a1-alkyl-2-acetyl ether (plasmalogen) causes blood-platelet aggregation and dilatation of blood vessels.

Figure 1.5

Structures of phosphatidic acid, phosphatidylcholine, phosphatidylethanolamine, and phosphatidylinositol

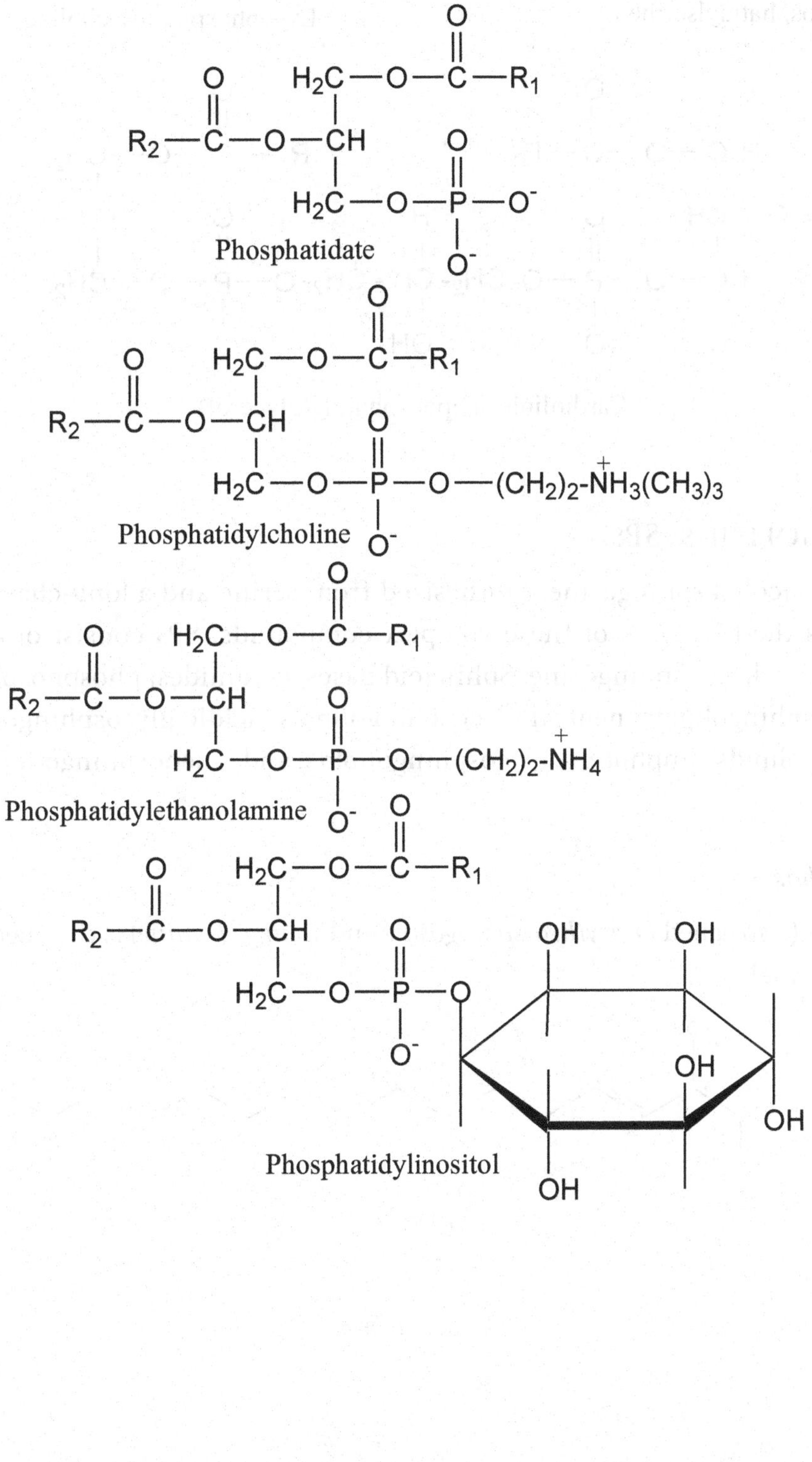

Figure 1.6

Structures of phosphatidylserine, lysophosphatidylcholine, and cardiolipin

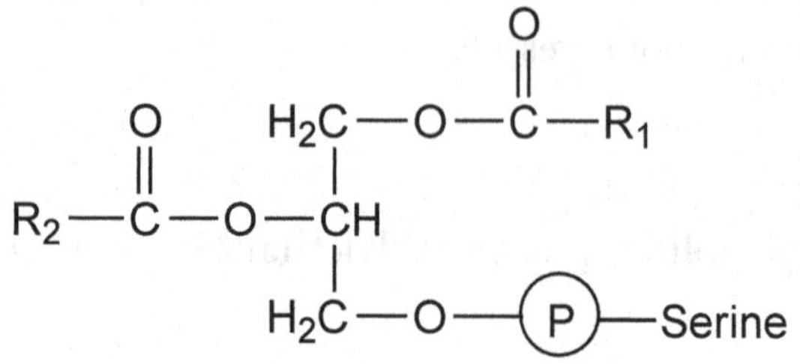

Phosphatidylserine

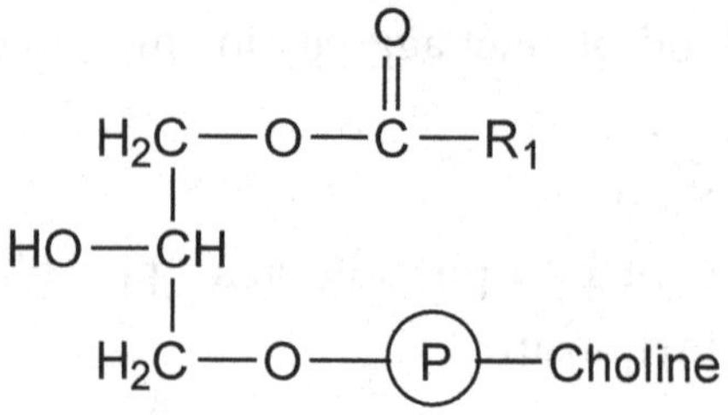

Lysophosphatidylcholine

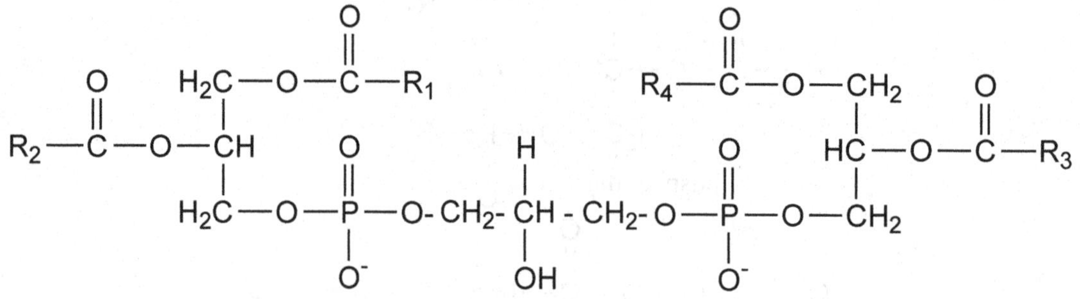

Cardiolipin (Diphosphatidylglycerol)

4) SPHINGOLIPIDS (SPs)

The amino alcohol sphingosine, synthesized from serine and a long-chain fatty acyl-CoA, forms the backbone of these complex compounds. SPs consist of a fatty acid, phosphoric acid, and sphingosine. Sphingoid bases, ceramides, phosphosphingolipids, phosphonosphingolipids, neutral glycosphingolipids, acidic glycosphingolipids, basic glycosphingolipids, amphoteric glycosphingolipids, and arsenosphingolipids are types of SPs.

Sphingoid bases

Sphingosine (also called *D-erythro-sphingosine* and *sphing-4-enine*) is the major sphingoid base in mammals.

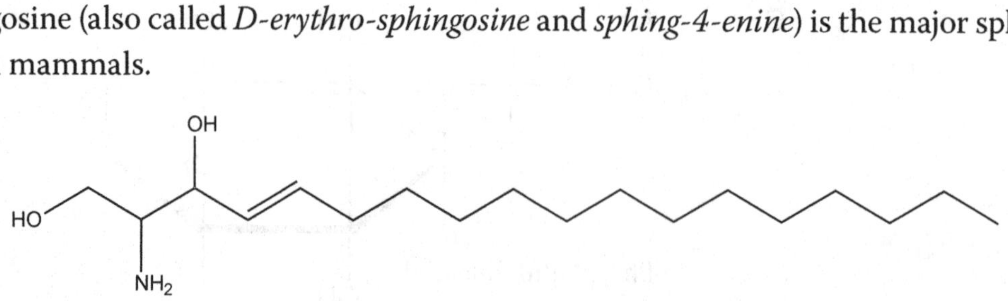

Sphingosine

Ceramides

Ceramides (N-acyl-sphingoid bases) are formed when a molecule of a fatty acid is attached to the amino alcohol sphingosine with an amide linkage. The fatty acids may be saturated or unsaturated with a chain length of 16 to 26 carbon atoms.

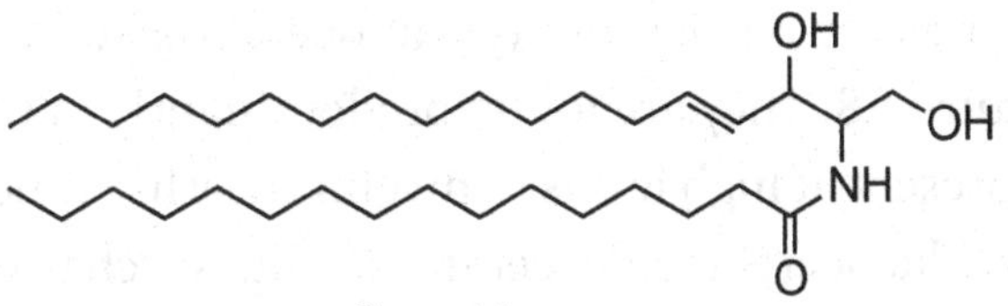

Ceramide

Sphingomyelins

The amino-alcohol sphingosine forms the backbone of sphingomyelins (ceramide phosphorylcholines), the major sphingophospholipids in mammals. These molecules contain a phosphorylcholine head group, a sphingosine, and a fatty acid, the last of which is attached to the amino group of sphingosine by an amide linkage to produce a ceramide. The alcohol group at carbon 1 of sphingosine esterifies to phosphorylcholine to produce sphingomyelin, an important constituent of the myelin of nerve fibres. Sphingomyelins play a key role in the formation of lipid rafts – lipid microdomains enriched in cholesterol involved in many cell processes, including membrane sorting and trafficking, signal transduction, and cell polarization. The structure of sphingomyelin is shown in figure 1.7.

Figure 1.7

The structure of plasmalogen and sphingomyelin.

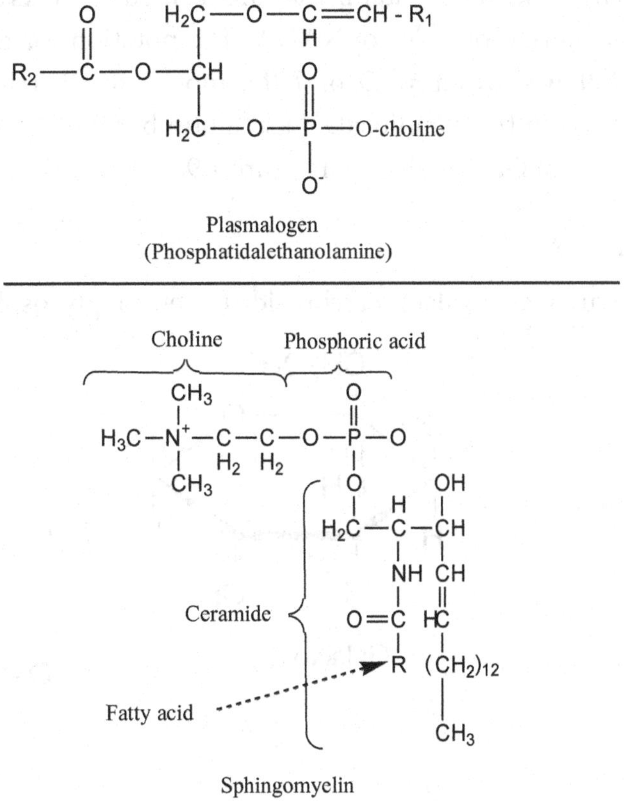

Plasmalogen
(Phosphatidalethanolamine)

Sphingomyelin

Glycosphingolipids

Almost all glycosphingolipids are derivatives of ceramides in which a fatty acid is attached to the amino alcohol sphingosine. Their polar head group is a monosaccharide or an oligosaccharide and is attached directly to the ceramide by an O-glycosidic bond. The carbohydrate moieties present in a glycosphingolipids determine their type.

Neutral glycosphingolipids

Cerebrosides are the simplest known neutral glycosphingolipids. They are ceramide monosaccharides that contain either a galactose (galactocerebroside) or a glucose (glucocerebroside), as shown in figure 1.8. They are found predominantly in the brain and peripheral nervous tissue and are present at high levels in myelin sheaths. When additional monosaccharides are attached to glucocerebroside, ceramide oligosaccharides are formed.

Acidic glycosphingolipids

These compounds are negatively charged at physiologic pH. The negative charge is provided by N-acetylneuraminic acid (NANA) in gangliosides and by sulphate groups in sulphatides.

Gangliosides

These are the most complex glycosphingolipids and are found primarily in ganglion cells of the central nervous system. They are formed from ceramide oligosaccharides and contain one or more molecules of NANA. The notation for these compounds is G (for ganglioside) followed by an M, D, or T (for mono-. di-, or tri-) to indicate the number of NANA molecules in the ganglioside. Last, a number indicates the subspecies of ganglioside. The structure of GM1 is shown in figure 1.9.

Figure 1.8

The structure of galactocerebroside (a neutral glycosphingolipid).

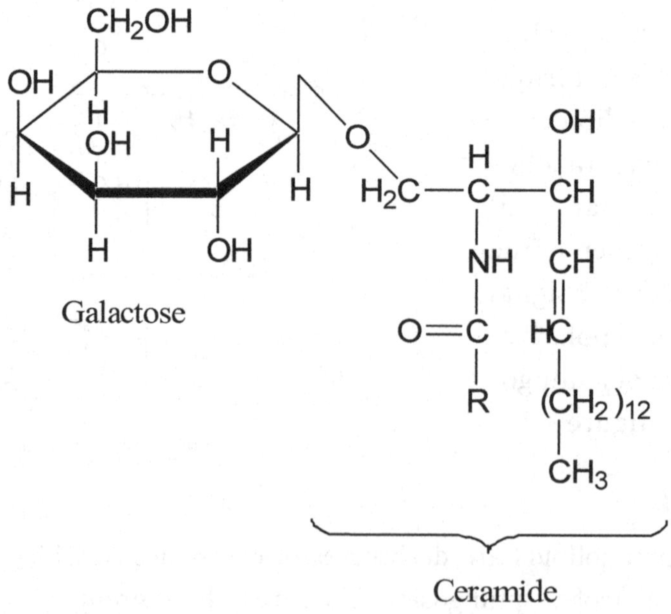

Sulphatides (sulphoglycosphingolipids)

These are cerebrosides that contain galactosyl residue that has been sulphated and is therefore negatively charged at physiologic pH.

Figure 1.9

The structure of ganglioside GM1

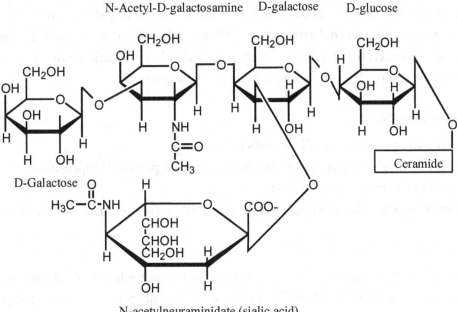

Major functions of sphingolipids

- Are essential components of all cell membranes in the body but in the greatest amounts in nerve cells. This is particularly true of sphingomyelins.
- Play a role in the regulation of cellular interactions, growth, and development.
- Are blood-group antigens and various embryonic antigens.
- Are surface receptors for cholera and diphtheria toxins and for some viruses.

5) STEROL LIPIDS (SLS)

Sterol lipids (SLs) are subdivided according to their biological function into sterols, cholesterol and its derivatives, steroids, secosteroids, bile acids and their derivatives, steroid conjugates, hopanoids, and others. SLs are essential components of cell membrane lipids. The steroids, all formed from the same fused four-ring core structure (figure 1.10), have biologic functions as hormones and signalling molecules. The C18 carbon steroids include the oestrogen family, and the C19 steroids include the androgens, such as testosterone and androsterone. The C21 steroids include the progestins, glucocorticoids, and mineralocorticoids. The secosteroids are characterized by the cleavage of the B ring of the steroid nucleus and include various forms of vitamin D. Bile acids and their derivatives are divided to four further subclasses, also according to the number of their carbon atoms: C24, C26, C27, and C28 bile acids, alcohols, and derivatives.

Cholesterol

Cholesterol has been extensively studied because of its link with cardiovascular disease. Deposits of cholesterol and cholesterol esters in the connective tissue of the arterial walls characterize atherosclerosis. Its carbon atoms are numbered as shown in figure 1.11, and it contains two angular methyl groups: the C19 methyl group attached to C10 and the C18 methyl group attached to C13. It also contains a branched C8-side chain at position 17, a double bond between C5 and C6, and a hydroxyl group at C3.

Functions of cholesterol

- Forms the precursor of all steroids in the body.
- Is an essential component of membranes and of plasma lipoproteins.
- Is a major constituent of gallstones.
- Form the precursor of vitamin D_3 (ergosterol is the precursor of vitamin D_2).

Steroids

Steroids are classed according to the number of carbons in their core skeletons, but they all contain the steroid nucleus (cyclopentanoperhydrophenanthrene [see figure 1.10]). The C18 subclass contains the oestrogen family; the C19, the androgens such as testosterone and androsterone; and the C21, whose molecules contain a C2 side chain at the C17 position, the progestogens, glucocorticoids, and mineralocorticoids.

Steroid hormone nomenclature

Cholesterol is the precursor of all steroid hormones: progestagens, glucocorticoids, mineralocorticoids, androgens, and oestrogens. All have a 17-carbon nucleus (shown in figure 1.12). Additional carbons can be added at C10 and C13 or as a side chain attached to C17. They differ in number and type of substituted groups, location of double bonds, and stereochemical configuration. The two angular methyl groups (C18 and C19) at positions 13 and 10 project in front of the ring system and serve as the point of reference. All substituents that are in the same plane as these groups are designated *cis*, or ß, and are represented by solid lines. In contrast, all substituents that are below the plane of the ring system are designated *trans*, or α, and are denoted by a dashed or dotted line. Double bonds are referred to by the number of the carbon that precedes them. These molecules' names depend on whether a hormone has one angular methyl group (estrane, 18 carbons), two angular methyl groups (androstane, 19 carbons), or 2 angular groups plus a 2-carbon side chain at C17 (pregnane, 21 carbons), as shown in figure 1.12.

Figure 1.10

The steroid nucleus.

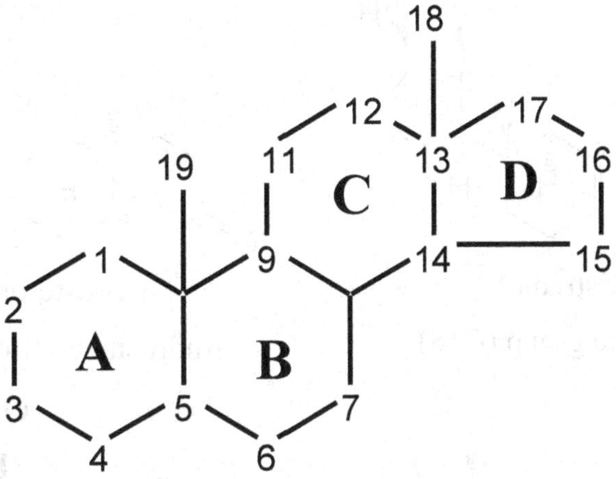

Secosteroids

The various forms of vitamin D make up this group of steroid. They are characterized by a cleavage of the B ring of the steroid nucleus. Figure 1.13 shows the structures of vitamin D_2 and vitamin D_3.

Figure 1.11

Cholesterol structure.

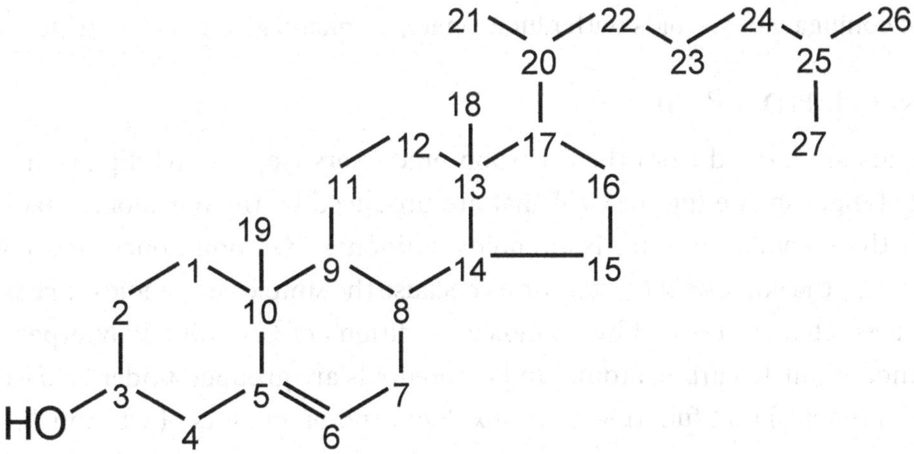

Bile acids

Cholesterol catabolism in the liver forms bile acids. The structure of chenodeoxycholic acid, which is representative of the C24 bile acids, alcohols, and derivatives, is shown in figure 1.14.

Figure 1.12

Basic steroid hormone structures

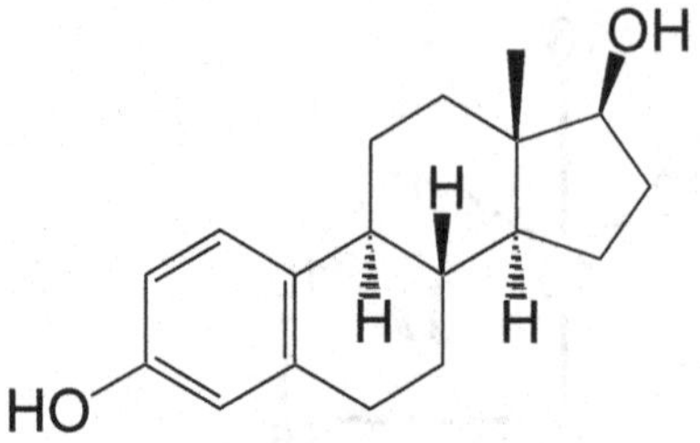

lEstradiol

Estrane group (C18)

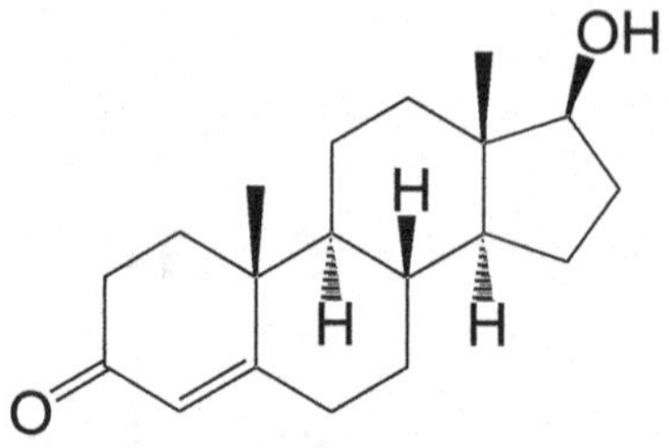

Testosterone

Androstane group (C19)

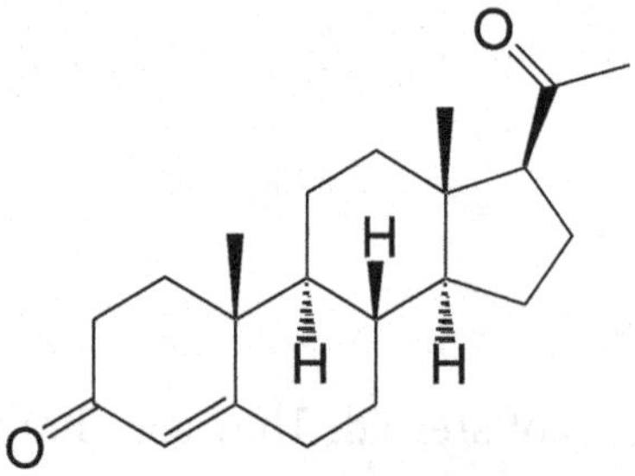

Progesterone

Pregnane group (C21)

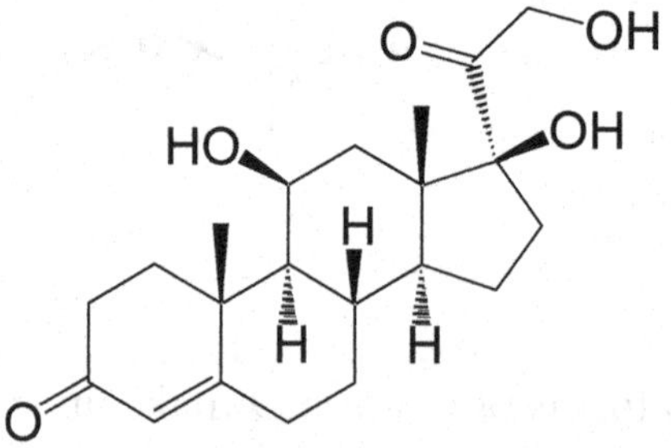

Cortisol

Pregnane group (C21)

Steroid conjugates

These are conjugate of steroids with glucuronate, sulphate, glycine, or taurine.

6) PRENOL LIPIDS (PRs)

Prenol lipids are formed from the 5-carbon precursors isopentenyl diphosphate and di-methylallyl diphosphate (figure 1.15) that are produced by the mevalonate pathway. The classes of these compounds are isoprenoids, quinones, hydroquinones, and polyprenols determined by the number of terpene units because the simple isoprenoids (linear alcohols, diphosphates, etc.) are formed by successive additions of C5 units. Polyterpenes usually contain more than 40 carbon atoms, and carotenoids are grouped under C20 isoprenoids (simple isoprenoids) that function as antioxidants and precursors of vitamin A.

Figure 1.13

The structures of vitamins D$_2$ and D$_3$.

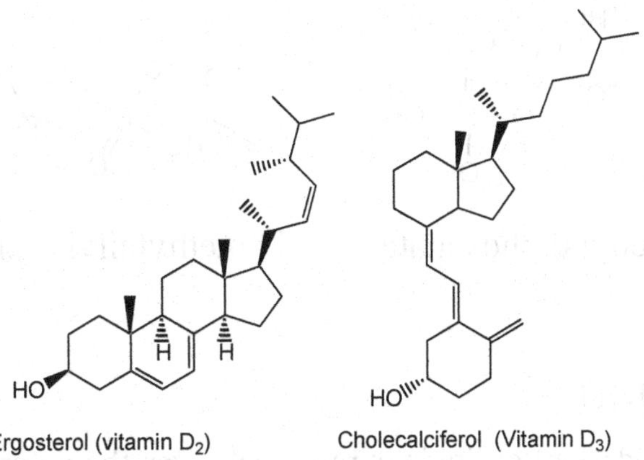

Ergosterol (vitamin D$_2$)   Cholecalciferol (Vitamin D$_3$)

7) SACCHAROLIPIDS (SLS)

SLs are formed when fatty acids are linked directly to a sugar backbone and can occur as glycan or as phosphorylated derivatives. The most common saccharolipids are the acylated glucosamine precursors of the lipid A component of the lipopolysaccharide found in gram-negative bacteria.

Figure 1.14

The structure of chenodeoxycholic acid.

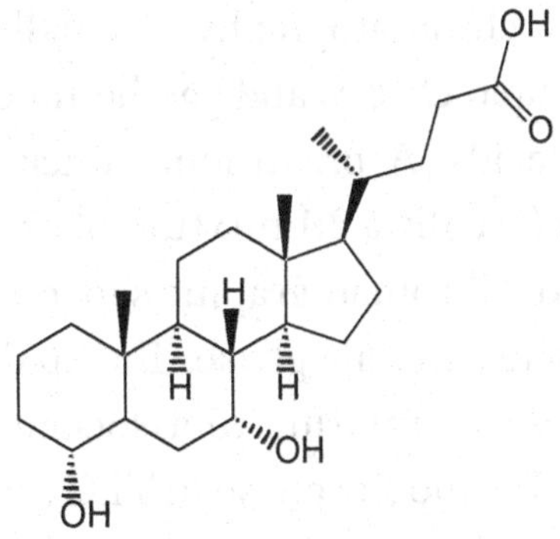

8) POLYKETIDES (PKS)

Polyketides structurally diverse but biosynthetically related family of natural products that includes medicinally important substances such as lovastatin (a cholesterol-lowering agent), erythromycin (an antibiotic), and FK506 (an immunosuppressant). Figure 1.16 shows the structure of lovastatin. PKs are synthesized by the polymerization of acetyl or propionyl units by classic, iterative, or multimodular enzymes with semiautonomous active sites that share mechanistic features with the fatty-acid synthases.

Figure 1.15

The structures of isopentenyl diphosphate and dimethylallyl diphosphate.

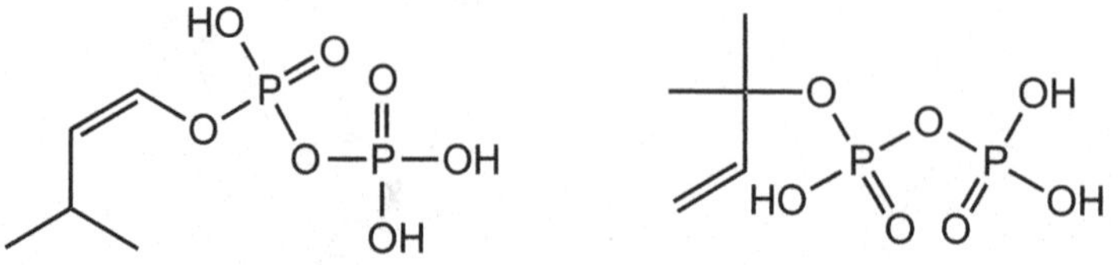

Isopentenyl diphosphate   Dimethylallyl diphosphate

LIPID PEROXIDATION

Peroxidation (auto-oxidation) of lipids is responsible for the deterioration of foods (rancidity) and damage to human tissue in vivo that may lead to cancer, inflammatory disease, atherosclerosis, and other effects. The peroxidation process is initiated by free radicals (ROO·, RO·, OH·) formed during peroxide formation from polyunsaturated fatty acids. For details, see chapter 10.

ANTIOXIDANTS

Naturally occurring antioxidants include vitamin E, vitamin C, ß-carotene, and glutathione. See chapter 10.

SEPARATION AND IDENTIFICATION OF LIPIDS

To separate and identify lipids, chromatography is usually used. Thin-layer chromatography (TLC) separates various lipid classes and gas-liquid chromatography (GLC) is used to separation individual fatty acids. A practitioner must extract lipids from biological samples using a solvent system (usually a 2:1 mixture of chloroform and methanol) before chromatography can be applied. Chromatographic separations partition molecules when they are between a stationary and a mobile phase. The lipid separation will depend on the relative tendencies of molecules in a mixture to associate more strongly with either the stationary or the mobile phase. The mobile phase in TLC is often a liquid, with the mixture being a suitable solvent system such as 80:20:2 hexane, diethyl ether, and formic acid. The stationary phase in GLC is normally an inert solid, such as silica gel or inert granules of ground firebrick, coated with a nonvolatile liquid, such as lubricating grease or silicon oil. The stationary phase in TLC is glass plates coated with silica gel. In GLC, molecules can be identified by comparison with the gas chromatographic pattern of a related standard mixture of known composition. In TLC, separated lipids can be visualized by exposure to iodine vapour, or by fluorescence (with dichlorofluorescein) and identify them by comparing them with a pattern of a standard mixture.

Figure 1.16

The structure of lovastatin.

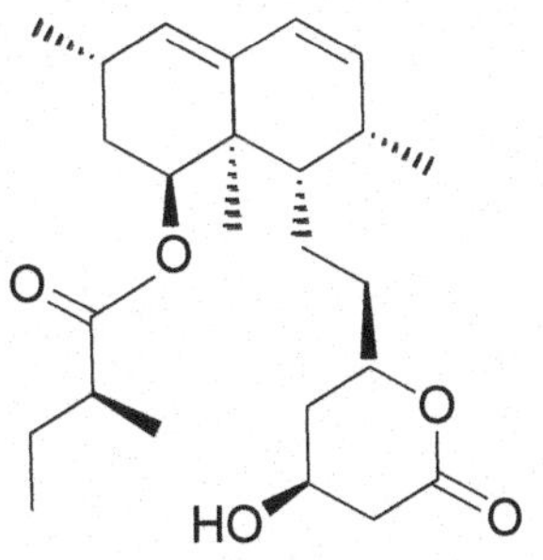

Lovastatin

CONCLUSION

- Lipids are a heterogeneous group of organic molecules important for cell-membrane structure and viscosity, nervous-tissue structure, physical protection of body organs, growth, energy metabolism, digestion, signal transduction, and regulatory and coenzyme functions.
- The eight major categories are fatty acyls, glycerolipids, glycerophospholipids, sphingolipids, saccharolipids, polyketides, prenols, and sterols. All of these compounds and their subclasses have important functions. Their biochemistry is the presented in the following chapters.

SELECTED READINGS

Brigelius-Flohe, R., and Traber, M. G. (1999). Vitamin E: Function and metabolism. *The FASEB Journal*, 13 (10): 1145–1155.

Burr, G, O.and Burr, M. M. (1930). The nature and role of the fatty acids essential in nutrition. J. Biol. Chem. 86: 587–621.

Fahy, E, Subramaniam, S, Brown, H. A. et al. (2005). A comprehensive classification system for lipids. *Lipid Research* 46: 839–862.

Gurr, A. I., and James, A. T. (1980). *Lipid biochemistry: An introduction*, 3rd ed. London: Chapman and Hall, 2–25.

Jones, G., Strugnell, S. A., and Deluca, H. F. (1998). Current understanding of the molecular actions of vitamin D. *Physiological. Reviews*, 78: 1193–1231.

Shearer, M. L., Bach, A., and Kohlmeier, M. (1996). Chemistry, nutritional sources, tissue distribution, and metabolism of vitamin K with special reference to bone health. *Journal of Nutrition* 126 (4 Suppl.):1181S–1186S.

von Linting, J. (2012). Metabolism of carotenoids and retinoids related to vision. *The Journal Biological Chemistry* 287 (3): 1627–1634.

Wong, F. T., and Khosla, C. (2012). Combinatorial biosynthesis of polyketides: A perspective. *Current Opinion on Chemical Biology*, 16 (1–2): 117–123.

Chapter 2: Digestion and Absorption of Dietary Lipids

Introduction

In Western countries, the average adult's dietary fat intake per day is about 60 to 150 g, of which approximately 90% is triacylglycerols (TAG). The remaining 10% is free fatty acids (FFAs), cholesterol esters (CE), phospholipids, and plant sterols. The saturated and unsaturated long-chain fatty acids make up bulk of the TAG Western adults consume, whereas medium-chain fatty acids (C6 to C14) found in plant oils (such as palm oil) and in animal fat (such as that in milk) constitute a relatively small percentage of the diet and are absorbed directly through the portal system by passive diffusion. Short-chain fatty acids, being water soluble, are freely absorbed.

Digestive enzymes break down most dietary fat from animal and plant sources into basic structures before the body absorbs them. Because lipids are hydrophobic, they must be emulsified for digestive enzymes to hydrolyze them. This emulsification is achieved by the action of bile salts, which increase the surface area of the fat droplets so that digestive enzymes can act on them more effectively. Dietary cholesterol esters, TAG, and phospholipids are degraded by pancreatic enzymes secreted under hormonal control, and the products of lipid digestion are then absorbed in the presence of bile salts by the brush-border membrane of the intestinal mucosal cells in the form of mixed micelles. TAG and cholesterol esters are resynthesized by the intestinal mucosal cells and finally incorporated into chylomicrons for secretion.

Digestion of Lipids in the Mouth and Stomach

In infants, digestion of dietary lipids begins in the stomach, where the process is catalyzed by acid-stable lipase (lingual lipase) secreted by Ebner's glands on the dorsal surface of the tongue. More specifically, this lipase catalyzes the release of short-chain fatty acids from position 3 of milk TAG. TAG that contain short- or medium-chain fatty acids, such as those found in milk fat, are digested by the gastric lipase produced by the chief cells of the gastric mucosa. Activity of this enzyme is markedly inhibited by protonated free fatty acids, which explains the limited hydrolysis of TAG under gastric conditions compared with the complete hydrolysis of TAG by pancreatic lipase in the duodenum. The short-chain (hydrophilic) fatty acids that are released are normally absorbed via the stomach wall into the portal vein.

EMULSIFICATION OF DIETARY LIPIDS

Emulsification of dietary lipids, which occurs in the duodenum, happens because of the detergent properties of bile salts and the mechanical mixing resulting from peristalsis. Enzymatic hydrolysis of TAG occurs only on the surface of a lipid droplet when the TAG is between the lipid droplet and the surrounding aqueous solution.

HORMONAL CONTROL OF LIPID DIGESTION

After a fat-rich meal, the pancreas secretes hydrolytic enzymes within pancreatic juices that degrade dietary lipids in the small intestine. Two intestinal hormones control this process: cholecystokinin (CCK; also called pancreozymin) and secretin. CCK stimulates the gallbladder to contract and release a micellar solution of bile acids, cholesterol, and phospholipids into the duodenum. It also stimulates the discharge and the continuous synthesis of pancreatic lipases. In addition, it decreases gastric motility, which slows the release of gastric contents into the small intestine. Secretin stimulates the pancreas to release a watery solution rich in bicarbonate, which helps to neutralize the pH of the intestinal contents, bringing it to the appropriate pH for the activity of the digestive enzymes.

DEGRADATION OF DIETARY LIPIDS

Triacylglycerol (TAG)

TAG is digested by pancreatic lipase, which catalyzes the removal of fatty acids at carbons 1 and 3 to produce 2-monoacylglycerol (2-MAG), an important end product of TAG digestion. The fatty acid at position 2 in the 2-MAG moiety requires isomerization to form 1-monoacylglycerol (1-MAG) before it is removed by pancreatic lipase. Less than one-fourth of dietary TAG is completely hydrolyzed to glycerol and free fatty acids (see figure 2.1). The pancreatic protein colipase helps to anchor and stabilize pancreatic lipase at the lipid-aqueous interface. Free fatty acids are then taken up from the intestinal lumen into the enterocytes by FATP4, a member of the fatty-acid transport-protein (FATP) family.

Figure 2.1

Enzymatic degradation of dietary TAG.

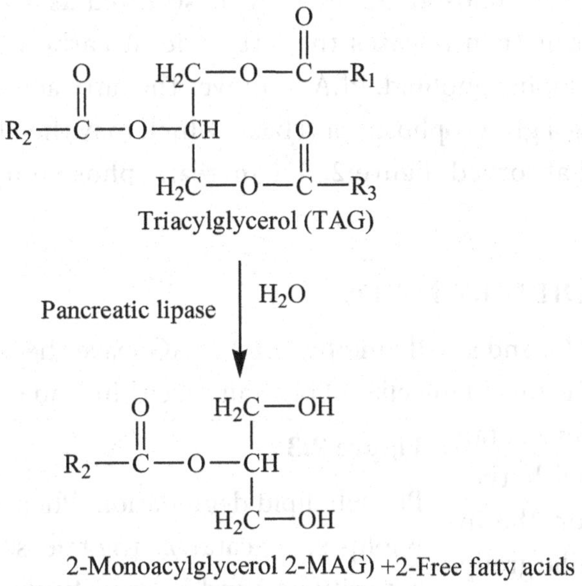

Triacylglycerol (TAG)

Pancreatic lipase | H_2O

2-Monoacylglycerol 2-MAG) +2-Free fatty acids

Cholesterol ester (CE)

In the small intestine, cholesterol esterase hydrolyzes cholesterol esters to produce free cholesterol and free fatty acids, which are absorbed from the intestine in their free form (figure 2.2). Bile salt concentrations within the range found in the intestinal lumen tend to block cholesterol ester formation and to promote cholesterol ester hydrolysis.

Figure 2.2

Cholesterol ester degradation.

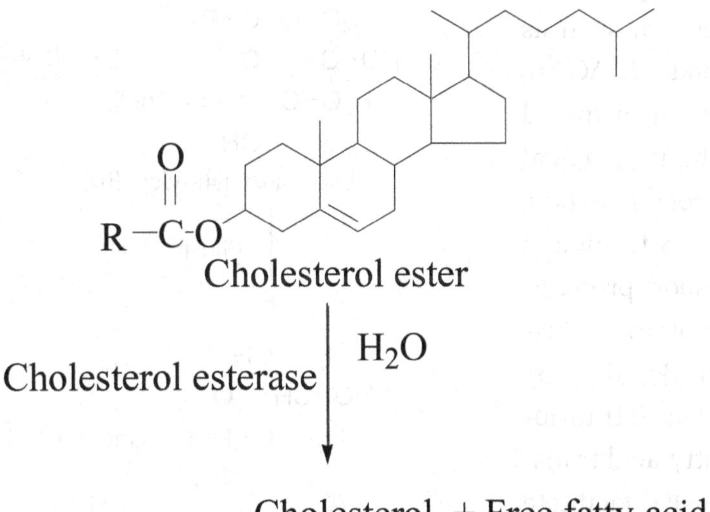

Cholesterol ester

Cholesterol esterase | H_2O

Cholesterol + Free fatty acid

Phospholipids

Phospholipase A_1 (PLA_1) and phospholipase A_2 (PLA_2) are pancreatic enzymes required for degradation of dietary phospholipids. PLA_2 is secreted as a zymogen and requires trypsin for activation. It then releases the fatty acid at carbon 2 of the phospholipid substrate, leaving a lysophospholipid. PLA_1 removes the fatty acid at carbon 1 of the lysophospholipid, leaving a glycerophosphoryl base, which may then be excreted in feces or further degraded and absorbed (figure 2.3). Pancreatic phospholipases require calcium ion as a cofactor.

ABSORPTION OF DIETARY LIPIDS

Free fatty acids, 2-MAG, and small amounts of 1-MAG leave the oil phase of lipid emulsion and diffuse into the mixed micelles of bile salts, lecithin, and cholesterol furnished by the bile. These micelles contain amphipathic lipids, with their hydrophobic groups on the inside and their hydrophilic groups on the outside, in contact within the aqueous environment of the intestinal lumen (see figure 2.4). Because micelles are water soluble, they allow the products of lipid digestion to be transported through the aqueous environment of the brush border of mucosal cells and absorbed into the intestinal epithelium. Products of lipid digestion, as opposed to dietary lipid digestion, such as cholesterol, FFA, and 2-MAG are absorbed in the form of mixed micelles through the apical membrane of enterocytes, where the uptake process is facilitated by a group of transport proteins. Among these proteins, cluster-differentiation 36 (CD36), also known as fatty-acid translocase (FAT), and fatty-acid transport protein 4 (FATP4) facilitate the transport of fatty acids, and

Figure 2.3

Phospholipid degradation: Phospholipase A_2 and phospholipase A_1 catalyze the release of the two fatty acids at positions 2 and 1, respectively, on the glycerol moiety of phosphatidylcholine.

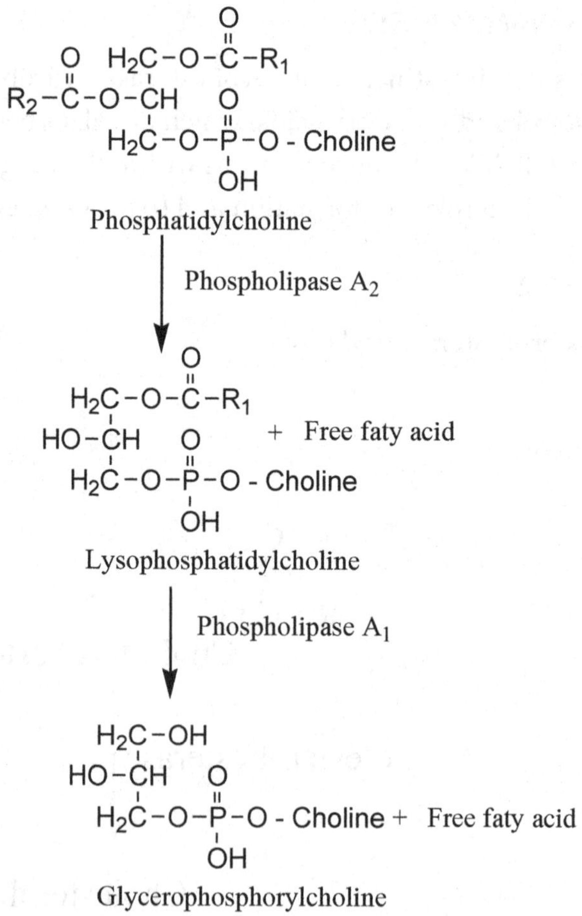

Niemann-Pick C1-like 1 (NPC1L1), a polytopic transmembrane protein, is essential for intestinal sterol absorption and the target of ezetimibe, a potent cholesterol-absorption inhibitor that lowers blood cholesterol.

....

Figure 2.4

The products of lipid digestion (e.g. cholesterol, FFA, 2-MAG) are absorbed in the form of mixed micelles through the apical membrane of enterocytes, where the uptake process of lipid products is facilitated by a group of transport proteins.

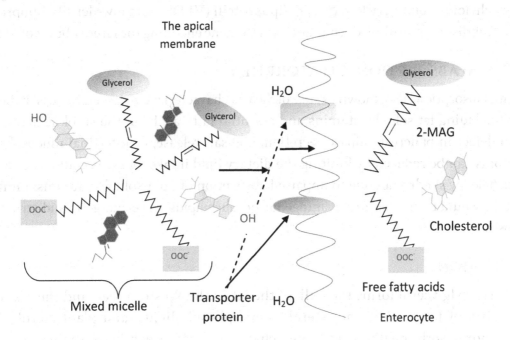

FORMATION OF CHYLOMICRONS (CMS)

In the enterocytes, absorbed 2-MAG, lysophospholipids, and cholesterol are reacylated with acyl-CoA (active fatty acid). TAG, phospholipids, and cholesterol ester is then regenerated in the smooth endoplasmic reticulum. In the presence of apolipoprotein B-48 (apo B-48) and microsomal triglyceride transfer protein (MTTP), resynthesized lipids and absorbed fat-soluble vitamins can assemble CMs. MTTP binds with high affinity to apo B and has lipid-transfer activity. Intestinal apo B-48 (2152 amino acid residue) is a polypeptide coded by the same gene that produces the much larger apolipoprotein B-100 (4536 amino acid residue) in hepatocytes. Thus, the human genome contains one apo B gene. In the hepatocytes, it is transcribed into a single mRNA sequence and is translated into apo B-100 as a single polypeptide of 4536 amino acids. However, in the intestine, the apo B mRNA is post-transcriptionally edited, resulting in the conversion of a glutamine codon into a stop codon. The edited mRNA is translated into apo B-48 as a single polypeptide of 2152 amino acids.

CMs form a milky fluid called chyle that is collected by the lymphatic vessels of the abdominal region and passed to the systemic blood through the thoracic duct. Short- and medium-chain fatty acids are released into the portal vein, where they are circulated by serum albumin to the liver but not activated.

ABETALIPOPROTEINEMIA

This rare autosomal recessive disorder is characterized by fat malabsorption. The fundamental biochemical defect is a complete absence of plasma apo B–containing lipoproteins, namely chylomicrons, very low-density lipoprotein (VLDL), and low-density lipoprotein (LDL). This disorder results from mutations in the gene encoding the large subunit of MTTP.

LIPID MALABSORPTION (STEATORRHEA)

Lipid malabsorption, also known as steatorrhea, is characterized by excessive loss of dietary lipids (including fat-soluble vitamins and essential fatty acids) in the stool. It may result from a defect in pancreatic lipase secretion, bile salt release, or intestinal mucosal cells. Symptoms can be reduced by limiting the dietary lipid intake, but this could lead to specific deficiency syndromes due to impaired absorption of lipid-soluble vitamins. Limiting fat intake would be further disadvantageous because lipids serve energetic and structural purposes.

CONCLUSION

- Triacylglycerol forms the bulk of the fat in the Western diet, and the rest consists of fatty acids, cholesterol esters, phospholipids, and plant sterols. The body absorbs medium- and short-chain fatty acids freely because they are water soluble.
- Under the control of gastrointestinal cholecystokinin and secretin, digestive enzymes are secreted into the small intestine, where they break down triacylglycerols into fatty acids and glycerol, and cholesterol esters into fatty acids and cholesterol. However, before lipid digestion can occur, lipids must be emulsified. This is achieved by action of bile salts, which increase the surface area of fat droplets so that digestive enzymes can effectively act on them. Then, in the presence of bile salts, hydrophobic products of lipid digestion are converted into mixed micelles, which are absorbed by the intestinal mucosal cells.
- Triacylglycerols and cholesterol esters are resynthesized by the intestinal mucosal cells and incorporated into chylomicrons for distribution to the body tissues.

SELECTED READINGS

Doege, H., and Sahl, A. (2006). Protein-mediated fatty acid uptake: Novel insights from in vivo models. *Physiology* 21 (4): 259–268.

Friedman, H. I. and Nylund, B. (1980). Intestinal fat digestion, absorption, and transport: A review. *The American Journal of Clinical Nutrition* 33 (5): 1108–1139.

Igbal, J., and Hussain, M. M. (2009). Intestinal lipid absorption. *American Journa of Physiology – Endocrinology and Metabolism* 296 (6): E1183–E1194.

Nelson, G. J., and Ackman, R. G. (1988). Absorption and transport of fat in mammals with emphasis on N-3 polyunsaturated fatty acids. *Lipids* 23: 1005–1014.

Vance, D. E., and Vance, J. (eds.). (1985). *Biochemistry of lipids and membranes.* Menlo Park, CA: Benjamin/Cummings.

CHAPTER 3: SYNTHESIS OF FATTY ACIDS

INTRODUCTION

Fatty acids synthesis occurs in the cytosol of most animal tissues when carbohydrates or proteins are present in high amounts in the diet. In humans, this mostly occurs in the liver, although it also takes place in lactating mammary glands, adipose tissue, and, to a lesser extent, the kidneys. Most fatty acids are synthesized from dietary glucose when it is converted to pyruvate in the cytosol by glycolysis. When pyruvate enters the mitochondria, it is converted to acetyl-CoA, which subsequently condenses with oxaloacetate to produce citrate. When citrate diffuses into the cytosol, it is cleaved by citrate lyase to yield oxaloacetate and acetyl-CoA, the immediate substrate for fatty-acid synthesis, which involves four enzyme systems: The extramitochondrial (cytosolic) system is responsible for biosynthesis of palmitate; the mitochondrial system, for chain elongation; and the microsomal systems, also for chain elongation and for desaturation of fatty acid. All four and the biosynthesis of eicosanoids are discussed in this chapter.

DE NOVO SYNTHESIS OF FATTY ACIDS

This type of synthesis, also referred to as lipogenesis, is responsible for palmitate synthesis from acetyl-CoA, which occurs in the cytosol. NADPH, ATP, Mn^{2+}, biotin, and bicarbonate (as source of CO_2), are required for lipogenesis.

Sources of acetyl-CoA

Acetyl-CoA, the building block of fatty acids, is formed in mitochondria from carbohydrates through glycolysis by the oxidative decarboxylation of pyruvate and by the degradation of some amino acids. Acetyl-CoA cannot cross the mitochondrial membrane, but the acetyl portion is transported to the cytosol in the form of citrate, which is produced by the condensation of oxaloacetate and acetyl-CoA within mitochondria. In the cytosol, citrate is cleaved by citrate lyase to produce acetyl-CoA and oxaloacetate. This process is enhanced when the mitochondrial citrate concentration is high and the ATP level is elevated (see figure 3.1).

Sources of NADPH

The pentose phosphate pathway is the most important source of NADPH. However, the $NADP^+$-dependent malate dehydrogenase (malic enzyme) and isocitrate dehydrogenase also contribute NADPH.

Figure 3.1

Transfer of acetyl-CoA from mitochondria to cytosol.

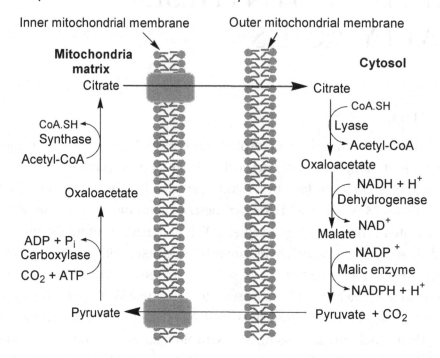

STEPS OF LIPOGENESIS

1. Formation of malonyl-CoA

Production of malonyl-CoA from acetyl-CoA is the initial step in fatty-acid synthesis. It is catalyzed by acetyl-CoA carboxylase, of which there are two forms: acetyl-CoA carboxylase 1 (ACC1) and acetyl-CoA carboxylase 2 (ACC2). In prokaryotes, ACC1 is composed of three distinct proteins: biotin carboxylase, biotin carboxyl carrier protein, and transcarboxylase. In the presence of ATP, biotin carboxylase transfers CO_2 from bicarbonate to the biotin carboxyl carrier protein to form the carboxybiotin derivative. Transcarboxylase catalyzes the transfer of the carboxyl group to acetyl-CoA to form malonyl-CoA (figure 3.2). In eukaryotes, however, three proteins are present as part of a single multifunctional protein, which is encoded by a single gene. ACC2 is associated with the outer mitochondrial membrane and forms malonyl-CoA, an inhibitor of the carnitine palmitoyl-transferase I (CPTI) activity, leading to inhibition of the transfer of the fatty-acyl group from the cytosol to the mitochondrial matrix, where β-oxidation of fatty acids occurs. The net result is reduced fatty-acid oxidation and increased fatty-acid and triacylglyceride (TAG) synthesis at the expense of glucose utilization.

Figure 3.2

Formation of malonyl-CoA from acetyl-CoA.

1. Biotin - enzyme + ATP + HCO_3^- $\longrightarrow$ CO_2 -biotin - enzyme + ADP + P_i

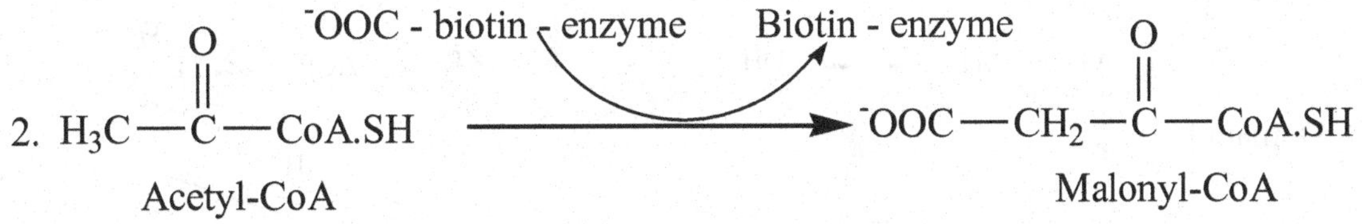

2. $H_3C - \underset{\overset{\|}{O}}{C} - CoA.SH$ $\xrightarrow{\qquad\qquad}$ $^-OOC - CH_2 - \underset{\overset{\|}{O}}{C} - CoA.SH$

   Acetyl-CoA $\qquad\qquad\qquad\qquad\qquad\qquad\qquad$ Malonyl-CoA

(over the arrow: ^-OOC - biotin - enzyme $\quad$ Biotin - enzyme)

2. *Production of palmitate*

In animals, a multicatalytic protein called fatty-acid synthase (FAS) catalyzes the synthesis of the long-chain saturated fatty acid palmitate in the presence of acetyl-CoA as primer, malonyl-CoA as the two-carbon donor, and NADPH as the reducing agent of the various intermediate products. In animals including humans, fatty-acid synthase is a dimer of two identical multifunctional polypeptides arranged in an antiparallel configuration, with each monomer having seven enzyme activities and an acyl carrier protein (ACP), to which the various generated intermediates are attached, on its 4'-phosphopantetheine group. In humans, the seven enzyme activities and the ACP are linked together covalently in a single polypeptide chain encoded by a single gene. Dimer formation is essential for these enzyme activities, which are organized from the N terminus to the C terminus as follows: β-ketoacyl synthase (KS), acetyl/malonyl transacylase (AT/MT), β-hydroxyacyl dehydratase (DH), enoyl reductase (ER), β-ketoacyl reductase (KR), ACP, and thioesterase (TE). The component activities are further grouped into three functional subdomains: domain I, containing KS, AT/MT, and DH; domain II, containing ER, KR, and ACP; and domain III, containing TE. In addition, domains I and II are separated by an interdomain (ID) polypeptide of 649 amino acids that have no known catalytic activity but are essential for dimer formation and generation of the active palmitate-synthesizing centre. Figure 3.3 shows a schematic diagram of a monomer of eukaryotic fatty-acid synthase.

Figure 3.3

Schematic diagram of monomer of eukaryotic fatty-acid synthase.

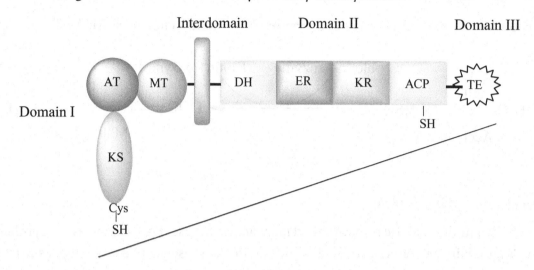

The reaction sequence that leads to the formation of palmitate from acetyl-CoA and malonyl-CoA can be summarized as follows:

1. A molecule of acetyl-CoA combines with the cysteine-SH group of KS by acetyl transacylase to produce acetyl-enzyme and CoA-SH.

2. A molecule of malonyl-CoA combines with the adjacent SH group of ACP of the other monomer by the action of malonyl transacylase to form malonyl-enzyme and CoA-SH.

3. Catalyzed by ß-ketoacyl synthase, the acetyl group attacks the methylene group of the malonyl residue and liberates CO_2 to form ß-ketoacyl enzyme (acetoacetyl enzyme). The energy of decarboxylation drives the reaction in the direction of synthesis.

4. The keto group of ß-ketoacyl enzyme is reduced to ß-hydroxybutyryl enzyme. This is catalyzed by ß-ketoacyl reductase, with NADPH as the reducing agent.

5. The ß-hydroxybutyryl enzyme is dehydrated to form crotonyl enzyme (2,3-unsaturated acyl enzyme) by the hydratase enzyme. Finally, crotonyl -enzyme (2 -*trans* -enoyl -ACP) is reduced to butyryl enzyme (acyl -enzyme) by enoyl reductase enzyme, and again, NADPH is the reductant. This completes the first turn in the spiral of fatty- acid synthesis.

In the second turn of the spiral, another molecule of malonyl-CoA combines with the SH group of ACP, displacing the saturated acyl residue onto the free cysteine-SH group. The elongation step is repeated 6 more times, with a new malonyl residue incorporated in each step until C16-acyl enzyme is formed. This intermediate is not a substrate for the condensing enzyme, so the synthetic process is terminated and free palmitate is released from FAS by the action of a thioesterase (deacylase). The reaction sequence that leads to the synthesis of fatty acids is shown in figure 3.4.

Figure 3.4

Reaction sequence in the synthesis of fatty acids.

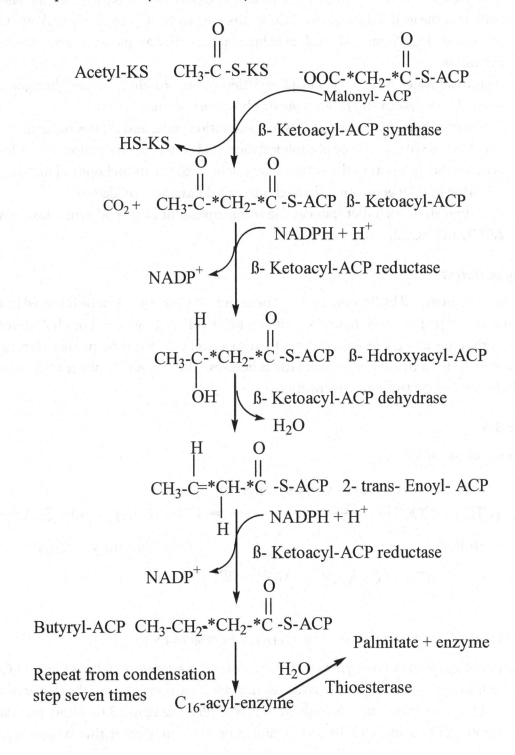

Regulation of lipogenesis

- Diet, especially a fat-free diet, induces the synthesis of ACC1 and ACC2 and increases their activities.
- Starvation or diabetes mellitus represses the expression of the carboxylase genes and decreases the activities of the ACC enzymes.
- Both ACC forms are allosterically activated by citrate, which is increased in concentration in the well-fed state. The ACC forms are inactivated by either malonyl-CoA or palmitoyl-CoA (the end product of lipogenesis), an example of negative-feedback inhibition.
- Insulin activates ACC by dephosphorylation of the protein, whereas glucagon and epinephrine deactivate the enzyme by phosphorylation.
- When metabolic fuel is low, as it is in a starvation state, and ATP is needed, ACC1 and ACC2 are turned off by phosphorylation, and consequently, malonyl-CoA levels are reduced, promoting the entry of fatty acids into the mitochondrial matrix and resulting in ATP generation through increased fatty-acid oxidation.
- A carbohydrate-rich diet induces the transcription of fatty-acid synthetase (FAS), ACC1, and ACC2.

Fate of palmitate

The palmitate produced by lipogenesis must be activated to acyl-CoA to be involved in any other metabolic pathway (see figure 3.5). It can be esterified to glycerol or cholesterol to form acylglycerols and cholesterol palmitate, respectively, or it can be further elongated, desaturated, or both by separate enzymatic processes. These can then form acylglycerols and cholesterol esters through esterfication.

· ·

Figure 3.5

Activation of palmitate.

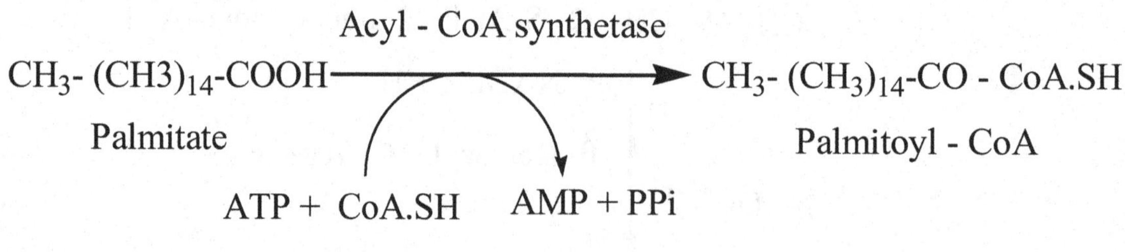

$$CH_3-(CH3)_{14}-COOH \xrightarrow{\text{Acyl - CoA synthetase}} CH_3-(CH_3)_{14}-CO-CoA.SH$$

Palmitate 　　　　　　　　　　　　　　　　　　　　Palmitoyl - CoA

ATP + CoA.SH　　　AMP + PPi

THE MICROSOMAL SYSTEM FOR CHAIN ELONGATION

Elongation of fatty acids takes place in the endoplasmic reticulum, where the acyl-CoA compounds are converted to acyl derivatives that are 2 carbons longer. The donor of the 2 carbons is malonyl-CoA, and the reductant is NADPH (see figure 3.6). Either saturated or unsaturated fatty acids with 10 carbon atoms or more may enter this system, which is inhibited by fasting. During myelination of nerve fibers, elongation of stearoyl-CoA is markedly increased in order to provide the C22 and C24 fatty acids required for the formation of sphingolipids.

Figure 3.6

The microsomal system for chain elongation of fatty acids.

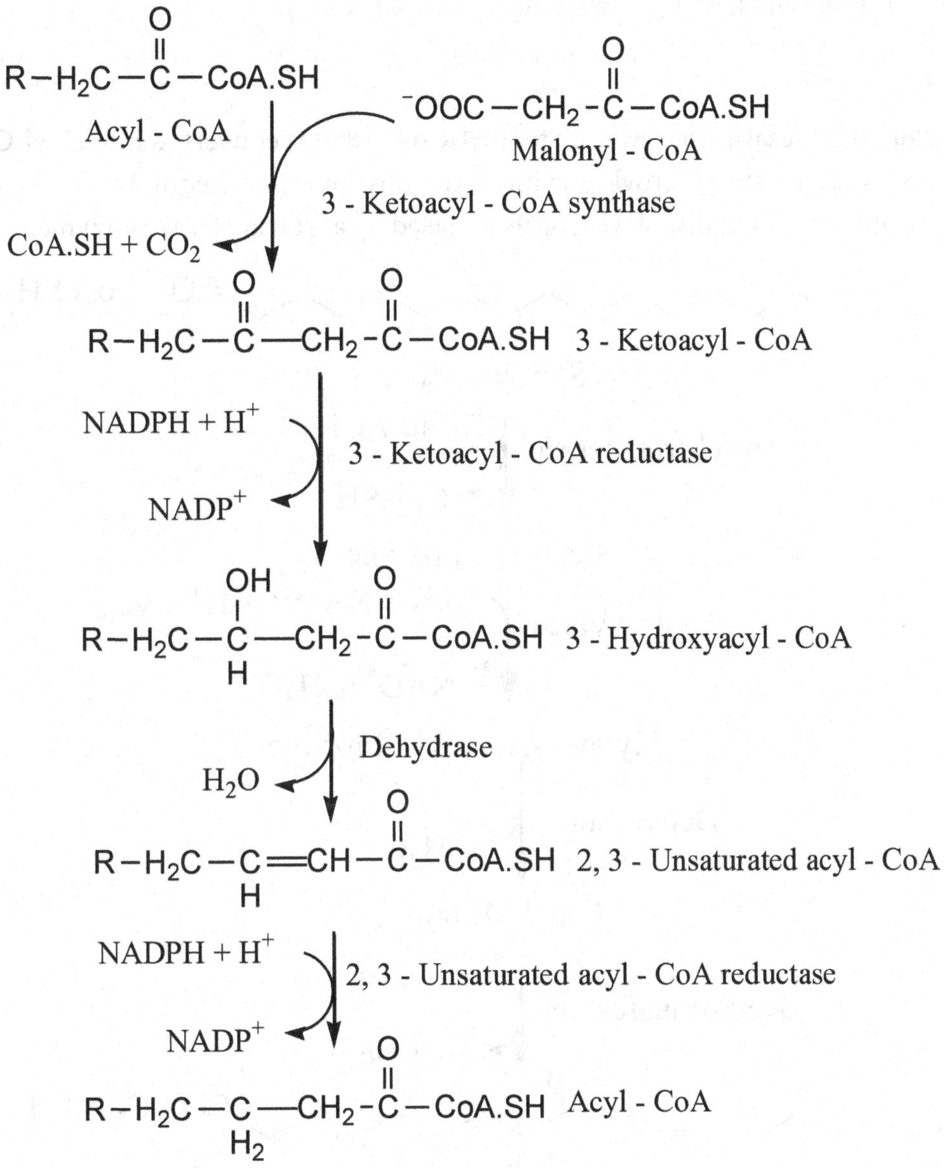

THE MITOCHONDRIAL SYSTEM FOR CHAIN ELONGATION

This system is responsible only for the elongation of fatty acids with moderate chain length. In mitochondria, saturated fatty acids (shorter than C16), are lengthened by successive additions of 2 carbons to the carboxyl-terminal end of the acyl residues.

BIOSYNTHESIS OF MONOUNSATURATED FATTY ACIDS

Monounsaturated fatty acids are synthesized from saturated fatty acids by several types of tissue, including liver tissue. Stearoyl-CoA (18:0) is the most frequent substrate for desaturation. The first double bond is nearly always introduced into a saturated fatty acid in the Δ^9 position. The microsomal Δ^9 desaturase system catalyzes conversion of palmitoyl-CoA or stearoyl-CoA to palmitoleyl-CoA or oleyl-CoA, respectively. This system requires oxygen, either NADH or NADPH, and cytochrome b_5 (see figure 3.7).

Figure 3.7

The microsomal Δ^9 desaturase system. Stearoyl-CoA (18:0) is converted into oleyl-CoA by Δ^9 desaturase system. The stearoyl-enzyme is then hydoxylated (requires, O_2 + NADH + H^+) and hydrated at Δ^9. Finally, oleyl-CoA is released by acyl transferase enzyme.

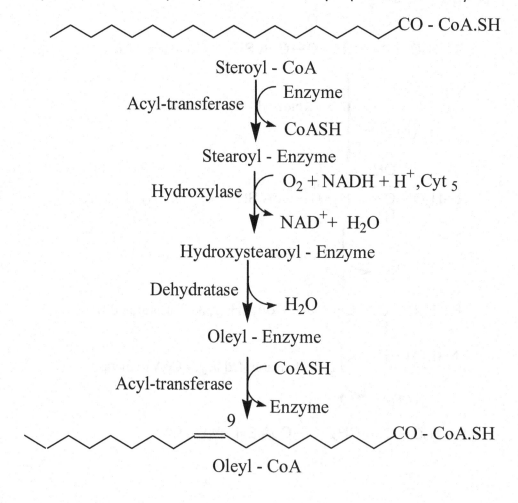

BIOSYNTHESIS OF POLYUNSATURATED FATTY ACIDS (PUFAS)

Eukaryotic cells produce a large variety of polyunsaturated fatty acids. Mammals can only desaturate fatty acids between the Δ^9 position and the carboxyl end of an acyl chain, whereas plants can desaturate them at positions Δ^9, Δ^{12}, and Δ^{15}. In plants, additional double bonds can be introduced between the existing double bond and the ω carbon. Mammals lack the desaturases required for the synthesis of either linoleic acid (ω-6) or α-linolenic acid (ω-3) and have a dietary requirement for these two fatty acids to accomplish the synthesis of the other members of the ω-6 and ω-3 families. Diets with a high ratio of polyunsaturated:saturated fatty acids lower serum cholesterol levels and therefore decrease the incidence of coronary heart disease. Figure 3.8 shows the conversion of linoleic acid to arachidonic acid. The reaction sequence of the pathway is described in the next section.

Conversion of linoleate to arachidonate

Linoleyl-CoA is converted to γ-linolenyl-CoA in a reaction catalyzed by Δ^6 desaturase in the presence of O_2 and NADH. The γ-linolenyl-CoA is then elongated to dihomo-γ-linolenyl-CoA by the microsomal elongase, which uses malonyl-CoA as the donor of the 2 carbon units. Finally, dihomo-γ-linolenyl-CoA is converted by Δ^5 desaturase to arachidonyl-CoA in the presence of O_2 and NADH.

Both fasting and low insulin levels inhibit the desaturation and chain-elongation processes.

BIOSYNTHESIS OF EICOSANOIDS

Eicosanoids are synthesized by nearly every cell in the body from arachidonic acid and some other C20 fatty acids and generally act at or close to the site of synthesis. Linoleic, arachidonic, and linolenic acid each synthesize a group of eicosanoids. Figure 3.9 shows the biosynthesis of eicosanoids from arachidonic acid. The arachidonic acid is released from membrane-bound phospholipids by phospholipase A_2. Prostanoids and leukotrienes are formed from arachidonate by the cyclooxygenase and the lipoxygenase pathways, respectively (figures 3.10 and 3.11).

The cyclooxygenase pathway

The first step in prostanoid synthesis is the consumption of two O_2 molecules catalyzed by prostaglandin endoperoxide synthase. This enzyme has two activities: that of fatty acid cyclooxygenase, which requires two molecules of O_2 for its activity, and that of peroxidase, which requires reduced glutathione. Figure 3.10 shows the pathway.

Arachidonate is converted to PGG_2 by cyclooxygenase. PGG_2 is then converted to PGH_2 (the precursor for prostanoids) by the peroxidase. Last, PGH_2 is converted to prostaglandins (PGD_2, PGE_2, and $PGF_2\alpha$), thromboxane A_2 (TXA_2), and prostacyclin (PGI_2).

Figure 3.8

Conversion of linoleic acid into arachidonic acid. Arachidonic acid is formed from linoleyl-CoA by desaturation, elongation, and a second elongation. The two desaturation steps are catalyzed by Δ^6 desaturase and Δ^5 desaturase in the presence of O_2 and $NADH + H^+$, separated by an elongation step catalyzed by elongase that uses malonyl-CoA as the 2-carbon donor.

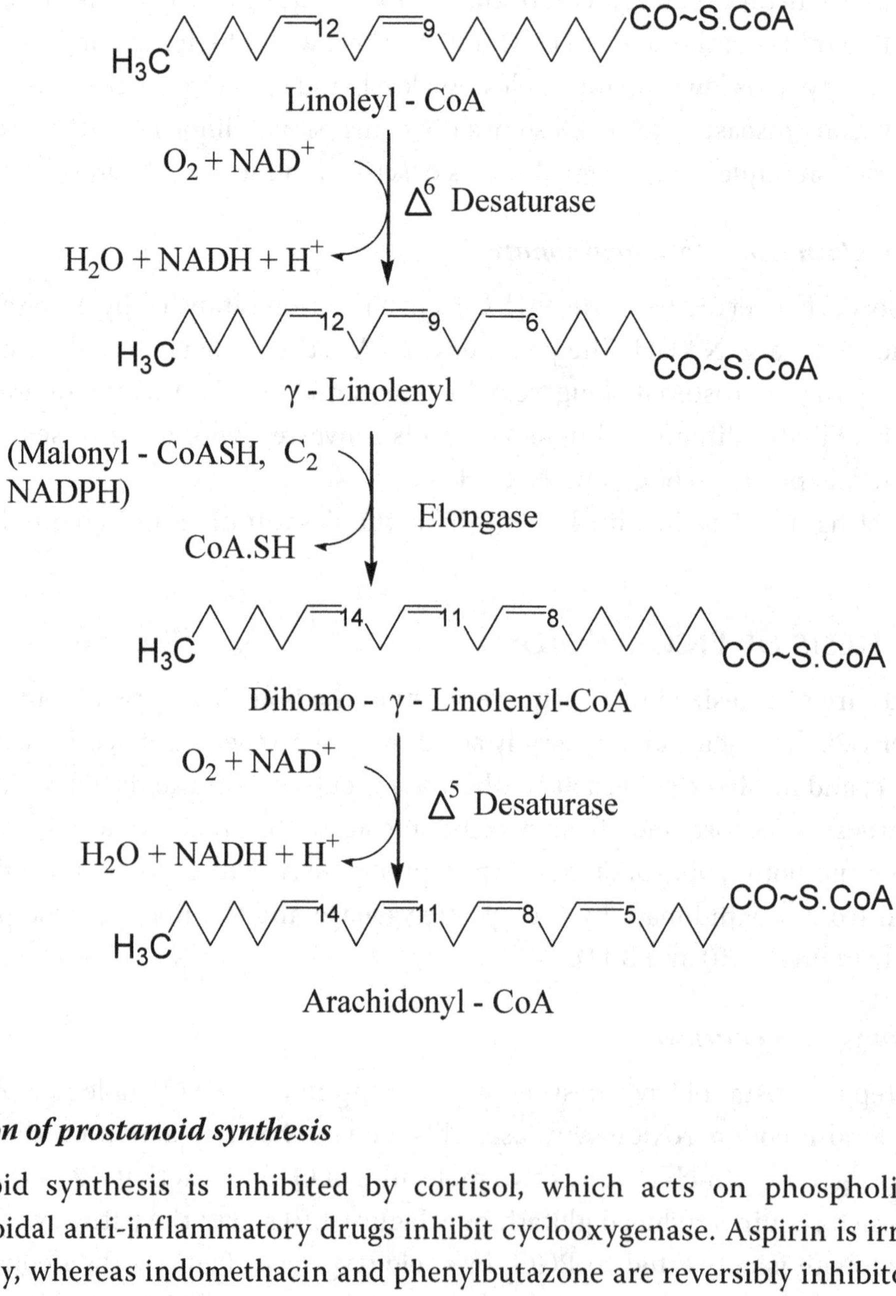

Inhibition of prostanoid synthesis

Prostanoid synthesis is inhibited by cortisol, which acts on phospholipase A_2. Nonsteroidal anti-inflammatory drugs inhibit cyclooxygenase. Aspirin is irreversibly inhibitory, whereas indomethacin and phenylbutazone are reversibly inhibitory.

Synthesis of leukotrienes is not affected by the nonsteroidal anti-inflammatory drugs.

Figure 3.9

Biosynthesis of eicosanoids from arachidonic acid. Eicosanoids are formed by the cyclooxygenase pathway and the lipoxygenase pathway. PG: prostaglandin, PGI: prostacyclin, TX: thromboxane, LT: leukotriene.

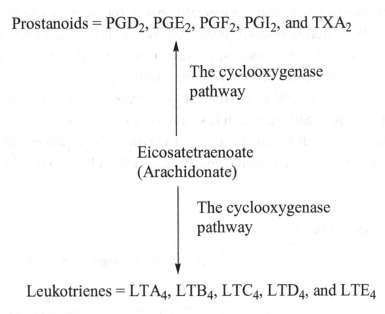

Prostanoids = PGD_2, PGE_2, PGF_2, PGI_2, and TXA_2

The cyclooxygenase pathway

Eicosatetraenoate
(Arachidonate)

The cyclooxygenase pathway

Leukotrienes = LTA_4, LTB_4, LTC_4, LTD_4, and LTE_4

The lipoxygenase pathway

Lipoxygenases (LOX) are non-heme iron-containing proteins that belong to a heterogeneous family of lipid-peroxidizing enzymes. These proteins are involved in the biosynthesis of the mediators of inflammation and classified according to the position of oxygen insertion into arachidonic acid as 5-, 8-, 12-, and 15-LOX. They insert oxygen at carbon 5, 8, 12, or 15 to produce 5S-, 8S-, 12S-, or 15S-hydroperoxyeicosatetraenoic acid (5-, 8-, 12-, or 15-HPETE), respectively, which can then be further reduced by glutathione peroxidase to the hydroxyl forms 5-, 8-, 12-, or 15-HETE, respectively.

Leukotrienes play important roles in innate responses and pathological roles in inflammatory diseases. They are formed from arachidonic acid by the 5-LOX pathway. The first committed step in the synthesis of leukotrienes is catalyzed by 5-LOX, which converts arachidonate into leukotriene A_4 (LTA_4) in the presence of the 5-LOX activating protein (FLAP). LTA_4 can be converted into the potent chemo-attractant leukotriene B_4 (LTB_4) by LTA_4 hydrolase or into the powerful bronchoconstrictor leukotriene C_4 (LTC_4) by LTC_4 synthase. Figure 3.11 shows the 5-LOX pathway.

Biological functions of eicosanoids

- TXA_2, synthesized by the platelets, causes vasoconstriction and stimulates platelet aggregation. In contrast, PGI_2, synthesized by blood-vessel walls, inhibits platelet aggregation and causes vasodilatation.
- PGE_2, formed by most tissues, especially that of the kidneys, causes vasodilatation and stimulates uterine smooth-muscle relaxation. It has been used to induce labor.
- $PGF_{2\alpha}$, produced by most tissues, causes vasoconstriction, contraction of smooth muscle, and stimulation of uterine contractions.
- Leukotrienes C_4, D_4, and E_4 are the components of the slow-reacting substance of anaphylaxis called SRS-A, which is secreted by mast cells.
- Leukotriene B_4 (LTB_4) causes increased chemotaxis of polymorpho-nuclear leukocytes, the release of lysosomal enzymes, and the adhesion of white blood cells.

Figure 3.10

The cyclooxygenase pathway.

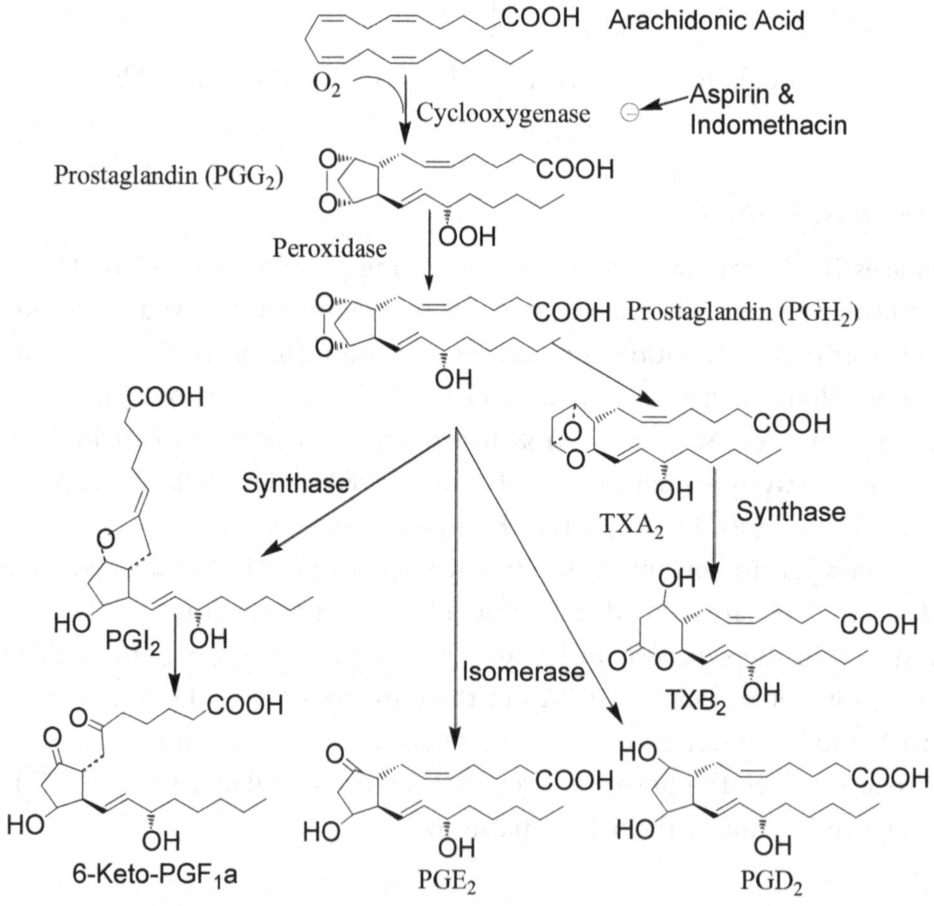

Figure 3.11

The 5-lipoxygenase pathway.

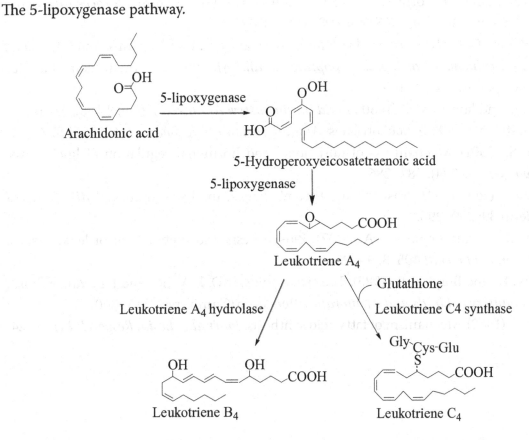

CONCLUSION

- Fatty acids are synthesized from acetyl-CoA in most animal tissues when carbohydrate or protein intake exceed bodily need. Fatty acid synthesis involves four enzyme systems: the de novo extramitochondrial (cytosolic) system for production of palmitate, the mitochondrial system for chain elongation, a microsomal system for chain elongation, and a microsomal system for the desaturation of fatty acid.

- Starvation, obesity, and diabetes mellitus affect fatty acid metabolism and reflect the important roles of insulin, glucagon, and adrenaline on this system.

- Prostanoids and leukotrienes (eicosanoids) are synthesized from arachidonic acid and other C20 fatty acids by the cyclooxygenase pathway and the lipoxygenase pathway.

SELECTED READINGS

Abu-Elheiga, L., Almarza-Ortega, D. B., Baldini, A., et al. (1997). Human acetyl-CoA carboxylase 2: Cloning, characterization, chromosomal mapping, and evidence for two isoforms. *The Journal of Biological Chemistry* 272: 10669–10677.

Chirala, S. S., Jayakumar, A., Gu, Z.-W., et al. (2001). Human fatty acid synthase: Role of interdomain in the formation of catalytically active synthase dimer. *Proceedings of the U.S. National Academy of Science* 98: 3104–3108.

Goodridge, A. G. (1991). Fatty acid synthesis in eucaryotes. In D. E. Vance and J. E. Vance (eds.), *Biochemistry of Lipids, Lipoproteins, and Membranes.* Amsterdam: Elsevier Science publishers, 11–139.

Gurr, A. I., and James, A. T. (1980). *Lipid biochemistry: An introduction*, 3rd ed., 25–67.

Hammarström, S. (1983). Leukotrienes. *Annual Review of Biochemistry* 52: 355–377.

Kersten, S. (2001). Mechanisms of nutritional and hormonal regulation of lipogenesis. *Embo. Reports* 2 (4): 282–286.

Moncada, S. (ed.). (1983). Prostacyclins, thromboxanes, and Leukotrienes. *British Medical Bulletin* 39: 209–295.

Murphy, R. C., and Gijon, M. A (2007). Biosynthesis and metabolism of leukotrienes. *Biochemical Journal* 405: 379–395.

Smith, W. L., and Borgeat, P. (1985). The eicosanoids. In D. E. Vance and J. E. Vance (Eds.), *Biochemistry of Lipids and Membranes.* Benjamin/Cummings, 325–360.

Wakil, S. J. (1960). Mechanism of fatty acid synthesis. *Journal. of Lipid. Research* 2 (1): 1–24.

Chapter 4: Oxidation Of Fatty Acids

Introduction

Fatty acids are oxidized along several pathways, with the primary one being the ß-oxidation pathway and secondary ones including α-oxidation, ω-oxidation, and peroxisomal oxidation of fatty acids. Heart and skeletal muscles derive a major source of energy from ß-oxidation, and ketone bodies are provided to extrahepatic tissues through hepatic ß-oxidation. In ß-oxidation, in the mitochondrial matrix, fatty acids are degraded by successive removal of 2-carbon fragments in the form of acetyl-CoA. To begin the process, fatty acids released from triacylglycerols (TAG) are transported to the mitochondrial matrix by a carnitine shuttle, which is regulated by malonyl-CoA formed in the committed step of lipogenesis. Thus, when fatty acids are synthesized, their oxidation is inhibited. Table 4.1 shows a comparison of fatty-acid synthesis and oxidation. Free fatty acids present in blood are bound to albumin and carried to most tissues for oxidation.

Lipolysis

Hydrolysis of TAG by specific adipose-tissue lipases must occur before FFAs and glycerol can be catabolized. TAG's complete hydrolysis into three free fatty acids and glycerol occurs by the action of adipocyte triglyceride lipase (ATGL), hormone-sensitive lipase (HSL), and monoacylglycerol lipase (MGL). ATGL is responsible for the first step in TAG mobilization, which generates diacylglycerol (DAG) and FFAs. HSL limits the rate of degradation of DAG into FAs and MAG. HSL can also hydrolyze ester bonds in sn-1 and sn-3 of TAG, but with lower substrate specificity than for DAG. MGL releases the third FA from MAG. (See figure 4.1.)

Table 4.1

Comparison between fatty-acid synthesis and degradation.

	Synthesis	Degradation
Major tissue site	Liver	Muscle & liver
Compartmentation	Cytosol	Mitochondria
Carrier	Acyl-carrier protein	Carnitine
Substrate	Acetyl-CoA	Palmitate
Product	Palmitate	Acetyl-CoA
Coenzyme	NADPH	NAD^+ and FAD
Nature of enzymes	Multicatalytic	Multienzyme
Key enzyme	Acetyl-CoA carboxylase	Carnitine-acyltransferase I
Activator	Citrate	Epinephrine
Inhibitor	Palmitoyl-CoA	Malonyl-CoA

Regulation of lipolysis

In human adipose tissue, catecholamines bind to β1 and β2 receptors and, with stimulatory Gs proteins, activate adenylate cyclase, leading to an increase in cAMP level and elevated activity of cAMP-dependent protein kinase-A (PKA). PKA-mediated phosphorlylation of lipolytic enzymes and lipid droplet-associated proteins increases TAG degradation (figure 4.2). Hormones known to stimulate PKA via Gs-protein-coupled receptors include glucagon, parathyroid hormone, thyrotropin, α-melanocyte-stimulating hormone, and adrenocorticotropin. Catecholamine (acting through α2-adrenergic receptors), adenosine (A1-adenosine receptor), prostaglandin (E2 receptor), neuropeptide Y receptor type-1 (NPY-1 receptor), and nicotinic acid (GPR109A receptor) antagonize lipolytic hormones by inhibiting adenylate cyclase activity. Insulin and insulin-like growth factor are the most potent inhibitory hormones of lipolysis. In addition, insulin stimulates phosphodiesterase, which breaks down cAMP to AMP, decreases lipolysis.

Fate of glycerol

Glycerol released by lipolysis is transported by the blood to the liver, where glycerol kinase converts it into glycerol-3-phosphate. Glycerol-3-phosphate formed in hepatocytes can be used in resynthesis of TAG or can be converted to dihydroxyacetone phosphate (DHAP), which can participate in glycolysis or gluconeogenesis. Glycerol kinase is also found in kidney, brown-adipose, intestine, and lactating mammary-gland tissue but not in adipocytes.

PATHWAYS OF FATTY ACID OXIDATION

ß-Oxidation

This pathway is the major pathway for catabolism of saturated fatty acids in mitochondria. In ß-oxidation, 2-carbon fragments in the form of acetyl-CoA are successively removed from acyl-CoA molecules. The C3 of a fatty acid is oxidized with subsequent cleavage between carbons 2 (α-carbon) and 3 (ß-carbon), hence the name ß-oxidation. Fatty acids must be activated before they can be catabolized by ß-oxidation.

Activation of fatty acids

Enzymes of ß-oxidation act on CoA esters, so to be oxidized, a fatty must first be converted into fatty acyl-CoA ester by acyl-CoA synthase (thiokinase) in the presence of ATP and CoA.SH, as shown in the following diagram:

$$1. \ R\text{-}CH_2\text{-}COOH + CoA.SH \xrightarrow{\text{Thiokinase}} R\text{-}CH_2\text{-}CO - CoA.SH$$

$$Mg^2 + ATP \qquad AMP + PPi$$

$$2. \ PP + H_2O \xrightarrow{\text{Pyrophosphatase}} 2 \ Pi$$

This activation process uses two ATP equivalents and one molecule of H_2O and is driven by the use of one high-energy phosphate bond. This is followed by the loss of an additional high-energy phosphate bond in the pyrophosphatase reaction.

Figure 4.1

Schematic diagram of lipolysis. ATGL is the primary enzyme responsible for TAG mobilization from adipocytes, generating DAG and FFA. HSL limits the rate of DAG degradation into FA and MAG. MGL releases the third FA from the glycerol backbone. ATGL: adipocyte triglyceride lipase, FFA: free fatty acid, DAG: diacylglycerol, HSL: hormone-sensitive lipase, MAG: monoacylglycerol, MGL: monoacylglycerol lipase, TAG: triacylglycerol.

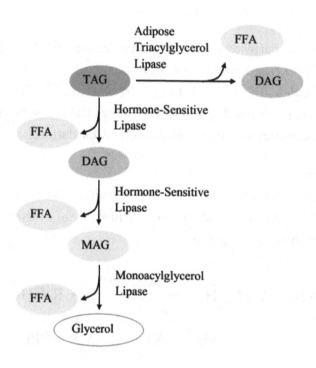

Role of carnitine in fatty-acid oxidation

Since the inner mitochondrial membrane is impermeable to acyl-CoA moieties, carnitine (ß-hydroxy-γ-trimethylammonium butyrate) is required for transport. The transport mechanism is called the carnitine shuttle (see figure 4.3).

The transport process can be summarized as follows: Acyl-CoA penetrates the outer membrane of mitochondria through pores. It reacts with carnitine to yield the acyl-carnitine derivative in a reaction catalyzed by carnitine-palmitoyltransferase I (CPT-1). Malonyl-CoA inhibits this enzyme, thus preventing fatty acids from entering into the mitochondrial matrix (this is the primary control point). In this way, when fatty-acid synthesis is in progress, malonyl-CoA prevents fatty-acid catabolism. The acyl-carnitine then is shuttled across the inner mitochondrial membrane by the carnitine-acyl-carnitine translocase (CACT), which acts as a membrane–carnitine acylcarnitine exchange transporter. Once inside the mitochondrial matrix, acyl-carnitine is converted back to its acyl-CoA derivative, the substrate for ß-oxidation. This reaction is catalyzed by carnitine-palmitoyltransferase II (CPT-2), which is attached to the inside of the inner mitochondrial membrane.

Figure 4.2

Schematic diagram of hormonal regulation of lipolysis. Hormonal binding to specific receptors activates adenylate cyclase, leading to an increase in cAMP levels and elevated activity of cAMP-dependent protein kinase-A (PKA). PKA-mediated phosphorlylation of lipolytic enzymes and lipid-droplet-associated proteins increases TAG degradation.

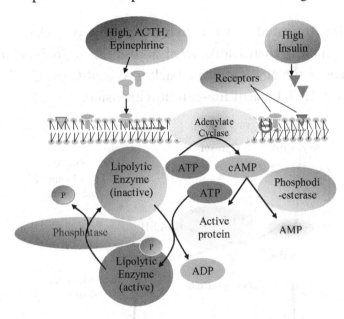

Carnitine deficiency

Carnitine deficiency results in impaired energy production from long-chain fatty acids, particularly during periods of fasting or stress. Primary carnitine deficiency is an autosomal recessive disorder that affects mitochondrial ß-oxidation and results from the lack of the organic cationic/carnitine transporter 2. It is characterized by hypoglycemia, muscle weakness, cardiomyopathy, and liver dysfunction. Secondary carnitine deficiency, such as medium-chain acyl-CoA dehydrogenase deficiency, results from defects in fatty-acid oxidation and is characterized by myopathy and hepatomegaly. It may also accompany alcoholic cirrhosis, renal failure, and chronic hemodialysis.

Carnitine palmitoyltransferase (CPT) deficiency

There are two CPT enzymes, CPT-1 (in the outer mitochondrial membrane) and CPT-2 (in the inner mitochondrial membrane), and three isoforms of CPT-1, each of which is encoded by a different gene – liver type (CPT-1A), muscle type (CPT-1B), and brain type (CPT-1C). Deficiency of only CPT-1A has been reported in humans.

CPT-1A deficiency is usually triggered by fasting or viral infection and is characterized by nonketogenic hypoglycemia, mild hyperammonemia, elevated plasma free fatty acids, and abnormal liver function tests. In children, it results in altered mental ability, stature and hepatomegaly.

CPT-2 deficiency is more common in adults but can occur during the neonatal period, with muscular involvement predominant. The neonatal form is characterized by respiratory distress, seizures, altered mental ability and stature, hepatomegaly, cardiomegaly, and cardiac arrhythmia.

Figure 4.3

The carnitine shuttle. Activated fatty acid in the form of acyl-CoA is converted by CPT-1 to acyl-carnitine, which is then transported across the mitochondrial membrane by CACT. CPT-2 converts the acyl-carnitine back into acyl-CoA. CACT: Carnitine-acyl-carnitine translocase, CPT-1: carnitine-palmitoyltransferase I, CPT-2: carnitine-palmitoyltransferase II.

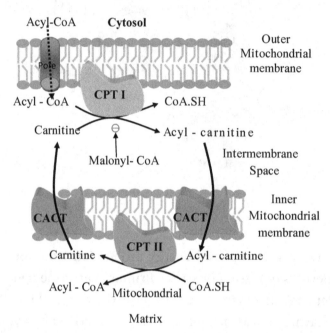

Reactions of ß-oxidation

Figure 4.4 outlines ß-oxidation of a saturated fatty acid. The process is catalyzed by enzymes found in the mitochondrial matrix known collectively as fatty acid oxidase. These enzymes catalyze oxidation of acyl-CoA to acetyl-CoA in a system that is coupled with the phosphorylation of ADP to ATP. The process uses NAD$^+$ and FAD as coenzymes and may be summarized as follows:

First, acyl-CoA is dehydrogenated by acyl-CoA dehydrogenase to produce Δ^2-transenoyl-CoA and FADH$_2$. The electrons from FADH$_2$ are transferred into the electron-transport chain, and the enoyl-CoA is then hydrated to form 3-hydroxyacyl-CoA in a reaction catalyzed by Δ^2-enoyl-CoA hydratase. The hydroxyl group is oxidized by L(3)hydroxyacyl-CoA dehydrogenase, with NAD$^+$ as a coenzyme, to produce 3-ketoacyl-CoA and NADH, and the electrons from NADH are transferred into the electron-transport chain. Finally, cleavage of the 3-ketoacyl-CoA by the thiol group of a second molecule of CoA yields acetyl-CoA and acyl-CoA shortened by 2 carbon atoms. This reaction is catalyzed by the 3-ketoacylthiolase. The acyl-CoA formed in this reaction then enters the oxidative pathway at reaction 1 (see figure 4.4). The enzymatic steps are repeated until butyryl-CoA is produced and is finally degraded to 2 acetyl-CoA molecules.

Figure 4.4

Process of ß-oxidation of saturated fatty acids. Long-chain saturated fatty acids (acyl-CoA) undergo repetition of reactions 1–4. Each turn of reactions results in the formation of acetyl-CoA and an acyl-CoA with two fewer carbons.

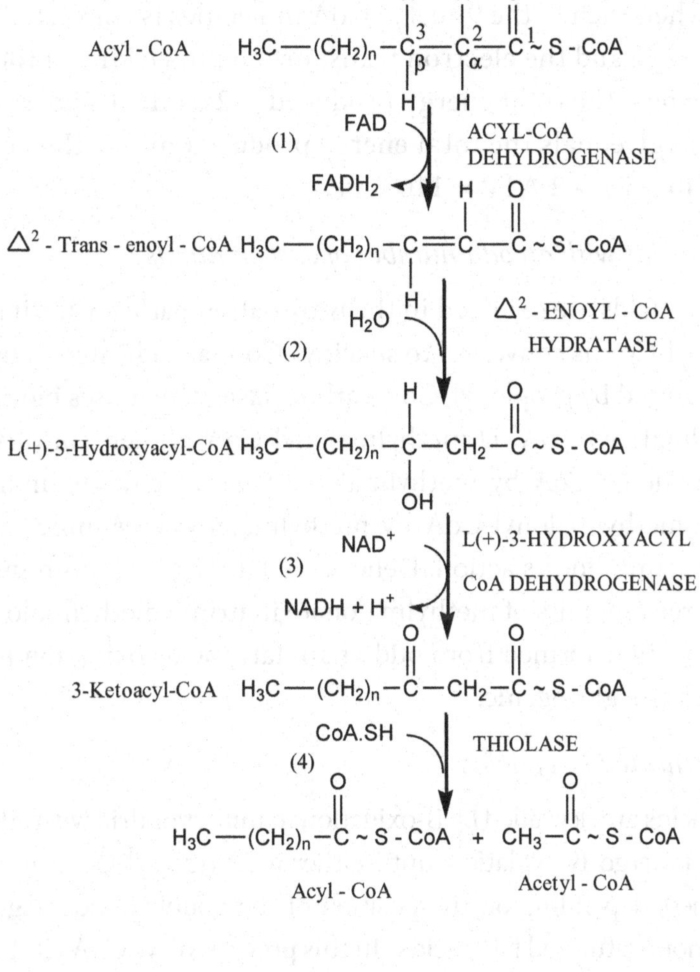

Energetics

Oxidation of fatty acids plays a crucial role in providing energy to body organs, particularly the liver, heart, and skeletal muscles, by generating large amounts of ATP. The oxidation of each molecule of NADH + H$^+$ and FADH$_2$ (generated during ß-oxidation) via the electron-transport chain results in the formation of 2.5 ATP molecules (theoretically 3 ATP) and 1.5 ATP molecules (theoretically 2 ATP), respectively. The complete oxidation of each molecule of acetyl-CoA (produced by ß-oxidation in the Krebs citric-acid cycle into CO_2 and H_2O) produces 10 ATP (theoretically 12 ATP) molecules.

The number of acetyl-CoA molecules produced can be easily calculated: The number of ß-oxidation turns equals the number of acetyl-CoA molecules formed minus 1. For example, the total energy generated by complete degradation of a molecule of stearic acid is 148 ATP. This is because stearic acid will be degraded into 9 acetyl-CoA molecules in 8 turns of ß-oxidation. In addition, 8 NADH+H^+ molecules are produced in these 8 turns of ß-oxidation, and each will provide 3 ATP when oxidized by the electron-transport chain, resulting in a total of 24 ATP molecules. Also, 8 $FADH_2$ are produced by the 8 turns of ß-oxidation, with each generating 2 ATP when oxidized by the electron-transport chain, resulting in a total of 16 ATP molecules. A further 12 ATP are formed when each of the 9 acetyl-CoA molecules is converted to CO_2 and H_2O in the citric-acid cycle and the electron transport chain, yielding a total of 108 ATP (9 x 12= 108 ATP). Thus, the total energy produced = 24 + 16 + 108 = 148 ATP. Finally, the net gain of energy equals the total energy produced minus the energy used in the activation step: 148 ATP – 2 ATP = 146 ATP.

Oxidation of fatty acids with an odd number of carbon atoms

The odd-chain fatty acids are oxidized in the ß-oxidation pathway until propionyl-CoA is produced. Propionyl-CoA is converted to succinyl-CoA, as indicated in figure 4.5. The initial reaction is catalyzed by propionyl-CoA carboxylase, which uses biotin as a coenzyme, to form D-methylmalonyl-CoA. D-methylmalonyl-CoA is then converted to its optical isomer, L-methylmalonyl-CoA by methylmalonyl-CoA racemase. Finally, succinyl-CoA is formed from L-methylmalonyl-CoA by methylmalonyl-CoA mutase, which requires deoxyadenosylcobalamin for its action. Deficiency of this last vitamin in humans leads to the excretion of large amounts of methylmalonate in urine (methylmalonic aciduria).

The propionyl residue formed from odd-chain fatty acids oxidation is the only known part of fatty acid that is glucogenic.

Oxidation of unsaturated fatty acids

Unsaturated fatty acids are degraded by ß-oxidation in mitochondria with additional enzymes. These fatty acids undergo ß-oxidation until either a Δ^3-*cis*-acyl-CoA or a Δ^4-*cis*-acyl-CoA compound is formed, depending on the position of the double bonds. Figure 4.6 shows the degradation of monounsaturated fatty acids . In this process, oleyl-CoA (18:1 Δ^9) is degraded by ß-oxidation until Δ^3-*cis*-dodecenoyl-CoA (12:1 Δ^3) is formed. Because this is not a substrate for acyl-CoA dehydrogenase, it is isomerized at the Δ^3-*cis* double bond into the Δ^2-*trans* by enoyl-CoA isomerase. The Δ^2-*trans*-dodecenoyl-CoA molecule is a normal substrate for the ß-oxidation enzyme enoyl-CoA hydratase, allowing the ß-oxidation pathway to be resumed. Since acyl-CoA dehydrogenase is not used, one fewer $FADH_2$ and two fewer ATP molecules are produced than in the ß-oxidation of the corresponding saturated fatty acid, such as stearoyl-CoA.

Figure 4.5

Conversion of propionyl-CoA to succinyl-CoA.

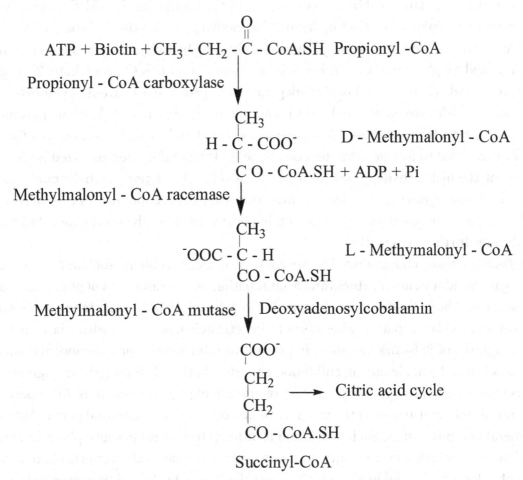

ATP + Biotin + CH₃ - CH₂ - C̈ - CoA.SH Propionyl -CoA

Propionyl - CoA carboxylase

CH₃
H - C - COO⁻   D - Methymalonyl - CoA
C O - CoA.SH + ADP + Pi

Methylmalonyl - CoA racemase

CH₃
⁻OOC - C - H   L - Methymalonyl - CoA
CO - CoA.SH

Methylmalonyl - CoA mutase | Deoxyadenosylcobalamin

COO⁻
CH₂   ⟶ Citric acid cycle
CH₂
CO - CoA.SH

Succinyl-CoA

Polyunsaturated fatty acids such as linoleyl-CoA are also undergo ß-oxidation, but with the help of enoyl-CoA isomerase and 2,4-dienoyl-CoA reductase. Figure 4.7 shows the degradation of linoleoyl-CoA. This polyunsaturated fatty acid undergoes three turns of ß-oxidation, as oleoyl-CoA does. The product is a Δ^3-*cis* unsaturated fatty acid, which is not a substrate for acyl-CoA dehydrogenase. However, the Δ^3-*cis* double bond is converted to the Δ^2-*trans* position by enoyl-CoA isomerase, allowing one additional cycle of ß-oxidation, resulting in the formation of Δ^4-*cis*-enoyl-CoA, which is then converted to Δ^2-*trans*, Δ^4-*cis*-dienoyl-CoA by acyl-CoA dehydrogenase. The Δ^2-*trans*, Δ^4-*cis*-dienoyl-CoA is converted by 2,4-dienoyl-CoA reductase to Δ^3-*trans* enoyl-CoA and then to the Δ^2-*trans* isomer by enoyl-CoA isomerase. This makes the complete oxidation of the remainder of the molecules by ß-oxidation possible.

α-Oxidation

This is a minor pathway in peroxisomes that oxidizes phytanic acid (3.7.11.15-tetramethylhexadecanoic acid), an important dietary component of animal fat and dairy products.

Phytanic acid is transported in plasma allied to very low-density lipoprotein (VLDL) and later in low-density lipoprotein (LDL). In addition, high-density lipoprotein (HDL) eliminates this branched fatty acid from tissue stores. In the human body, oxidation starts with the formation of a phytanoyl-CoA molecule. The methyl group on the ß-carbon of phytanic acid blocks the ß-oxidation pathway. This is resolved by hydroxylation of the α-carbon of phytanic acid by phytanoyl-CoA α-hydroxylase (which requires O_2, ascorbate, 2-oxoglutarate, iron, and ATP/GTP and is Mg^{2+}-dependent) to produce α-hydroxyphytanoyl-CoA. Subsequently, α-hydroxyphytanoyl-CoA is cleaved in the presence of thiamine pyrophosphate (TPP) by α-hydroxyacyl-CoA lyase into pristanal and formyl-CoA. Formyl-CoA is further converted to formic acid and coenzyme A. Pristanal is then oxidized by NAD^+-dependent aldehyde dehydrogenase to pristanic acid (2,6,10,14-tetramethylpentadecanoic acid), which undergoes three cycles of ß-oxidation after conversion into its CoA derivative (see figure 4.8). The products of ß-oxidation in peroxisomes are then transported to mitochondria for further degradation.

Refsum's disease, characterized by severe neurological problems such as tremors, unsteady gait, blindness, and deafness and accumulation of large amounts of phytanic acid in the tissues and body fluids, occurs in individuals who are unable to metabolize phytanic acid by the α-oxidation pathway because of a genetic deficiency of α-hydroxylase enzyme. Early diagnosis of Refsum's disease is important to minimize its progression, and such a diagnosis is usually made during childhood or early adulthood when retinitis pigmentosa results in apparent visual problems. Dietary restriction of phytanic acid is useful in preventing acute attacks and arresting the progression impairment of organs and particularly the peripheral nervous tissues. Such restriction is difficult to achieve because phytanic acid is found in meat, pelagic fish, and dairy products, and the human body converts phytol, a side chain of chlorophyll found in green leafy vegetables, to phytanic acid. However, long-term adherence to diets low in phytanic acid and phytol can lower plasma phytanic acid levels, which may be enhanced by serial plasma exchange to prevent development or progression of neuropathy, ataxia, cardiac arrhythmias, and ichthyosis.

ω-Oxidation

Under normal physiological conditions, the microsomal ω-oxidation pathway is minor; however, its activity is increased during fasting and starvation. It starts with hydroxylation of the methyl terminal (ω-carbon) to generate ω-hydroxy fatty acid by microsomal enzymes of the cytochrome P450 4A (CYP4A) and CYT F4 families (figure 4.9), which hydroxylate fatty acids of chain length C10 to C26 in the presence of NADPH and molecular O_2. The ω-hydroxylase CYP4A11 has a broad substrate spectrum and acts on lauric acid, myristic acid, palmitic acid, oleic acid, and arachidonic acid. CYP4F2 is the primary arachidonic acid ω-hydroxylase in human liver and kidney tissue and has a higher specificity for arachidonic acid than ω-hydroxylase CYP4A11. CYP4F2 ω-hydroxylates LTB4 in liver and the phytyl tails of tocopherols and tocotrienols (compounds that are collectively called vitamin E).

The ω-hydroxy fatty acid is further oxidized by NAD$^+$-dependent alcohol dehydrogenase into an ω-oxo fatty acid, which is then converted into the dicarboxylic acid by an aldehyde dehydrogenase that requires NAD$^+$. The dicarboxylic acid may be either excreted into urine or degraded from either end by the ß-oxidation pathway in peroxisomes or mitochondria.

Peroxisomal ß-oxidation

Long-chain and very-long-chain acyl-CoA are oxidized in peroxisomes by a modified form of ß-oxidation that shortens the acyl-CoA. The pathway proceeds only as far as C4 and C6 acyl-CoA since butyryl-CoA and hexanoyl-CoA are not substrates for the pathway. Instead, they are further degraded by ß-oxidation in mitochondria. Peroxisomal ß -oxidation involves desaturation, hydration of the α-enoyl-CoA, dehydrogenation of ß-hydroxyacyl-CoA, and thiolytic cleavage of ß-oxoacyl-CoA. The first oxidation, catalyzed by acyl-CoA oxidase, is not coupled to ATP synthesis. Instead, the high-potential electrons are transferred to O_2 to form H_2O_2, and catalase in peroxisomes converts the latter into H_2O and O_2.

DISEASES RESULTING FROM IMPAIRED OXIDATION OF FATTY ACIDS

Jamaican vomiting sickness

The unripe fruit of the akee tree contains the toxic, unusual amino acid hypoglycin A (usually referred to as hypoglycin) in the fruit and seed, and hypoglycin B, a γ-glutamyl derivative of hypoglycin, in the seed that is less toxic. When the fruit is eaten, hypoglycin binds irreversibly to coenzyme A, carnitine, and carnitine acyltransferase I and II, reducing their bioavailability and inhibiting ß-oxidation. The condition manifests in vomiting, seizures, and fatal hypoglycemia. Ripening of the fruit removes the toxic hypoglycin A, making it safe for consumption. Figure 4.10 shows the ackee fruit.

Dicarboxylic aciduria

This condition is characterized by the excretion of C6 to C10 ω-dicarboxylic acids and by nonketogenic hypoglycaemia. It results from lack of mitochondrial medium-chain acyl-CoA dehydrogenase.

Zellweger's syndrome

This syndrome, also called cerebrohepatorenal syndrome, is a rare and severe congenital disorder caused by a marked reduction in or absence of functional peroxisomes in all tissues. It is characterized by accumulation of C26 to C38 polyenoic fatty acids in brain tissue. Infants with Zellweger's syndrome experience a mild condition, with weak muscle tone (hypotonia), feeding problems, hearing loss, vision loss, and seizures.

X-linked adrenoleukodystrophy (ALD)

ALD occurs primarily in males and is characterized by the accumulation of very-long-chain fatty acids, especially in the adrenal cortex and the nervous system. This results from a defect in the peroxisomal ß-oxidation of very-long-chain fatty acids in peroxisomes caused by a mutation in the gene that encodes the peroxisomal transmembrane protein ALDP (X-ALD protein), which transports very-long-chain fatty acids from cytosol into peroxisomes. Those with this disease suffer deterioration of myelin sheaths around nerve fibers, which causes neurologic problems, and adrenal cortical malfunction, which causes Addison's disease. Adrenal-hormone replacement therapy is effective in correcting the adrenal deficiency.

Figure 4.6

Oxidation of oleoyl-CoA in mitochondria.

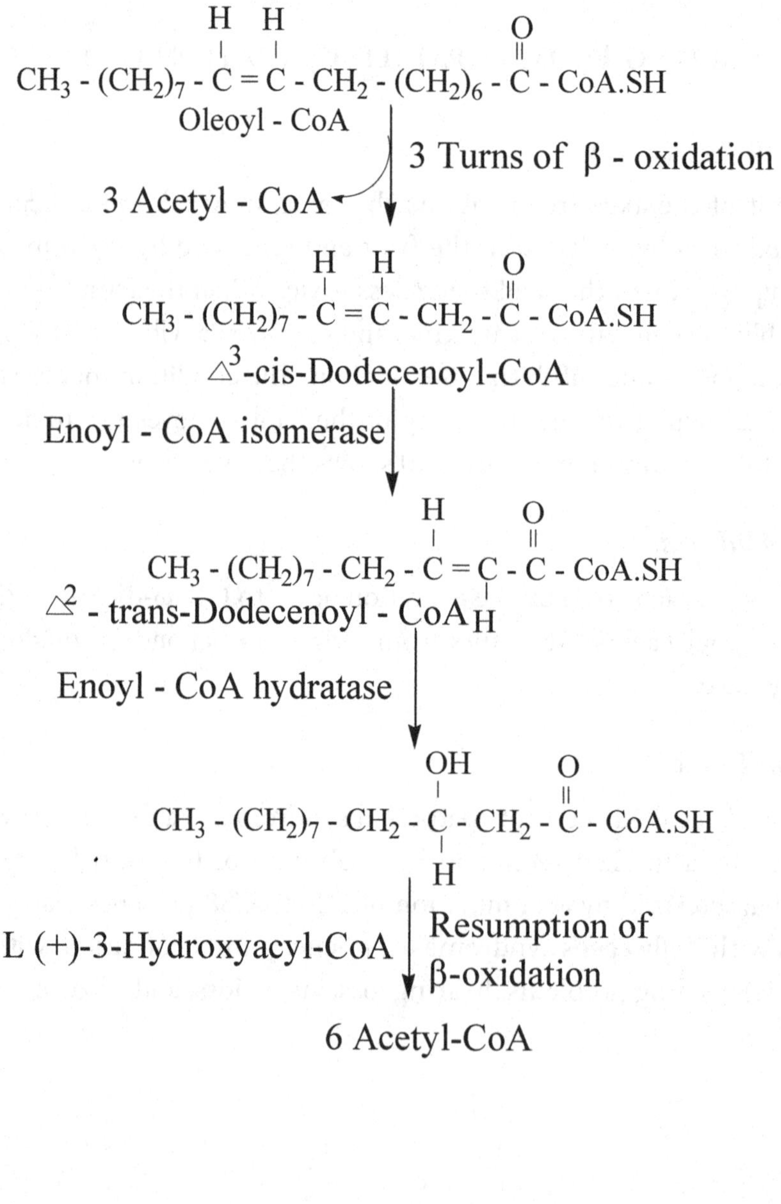

$$CH_3 - (CH_2)_7 - \overset{\overset{\displaystyle H}{|}}{C} = \overset{\overset{\displaystyle H}{|}}{C} - CH_2 - (CH_2)_6 - \overset{\overset{\displaystyle O}{||}}{C} - CoA.SH$$

Oleoyl - CoA

3 Turns of β - oxidation

3 Acetyl - CoA

$$CH_3 - (CH_2)_7 - \overset{\overset{\displaystyle H}{|}}{C} = \overset{\overset{\displaystyle H}{|}}{C} - CH_2 - \overset{\overset{\displaystyle O}{||}}{C} - CoA.SH$$

$\triangle^3$-cis-Dodecenoyl-CoA

Enoyl - CoA isomerase

$$CH_3 - (CH_2)_7 - CH_2 - \overset{\overset{\displaystyle H}{|}}{C} = \underset{\underset{\displaystyle H}{|}}{C} - \overset{\overset{\displaystyle O}{||}}{C} - CoA.SH$$

$\triangle^2$ - trans-Dodecenoyl - CoA

Enoyl - CoA hydratase

$$CH_3 - (CH_2)_7 - CH_2 - \underset{\underset{\displaystyle H}{|}}{\overset{\overset{\displaystyle OH}{|}}{C}} - CH_2 - \overset{\overset{\displaystyle O}{||}}{C} - CoA.SH$$

L (+)-3-Hydroxyacyl-CoA

Resumption of β-oxidation

6 Acetyl-CoA

Figure 4.7

The ß-oxidation of linoleoyl-CoA.

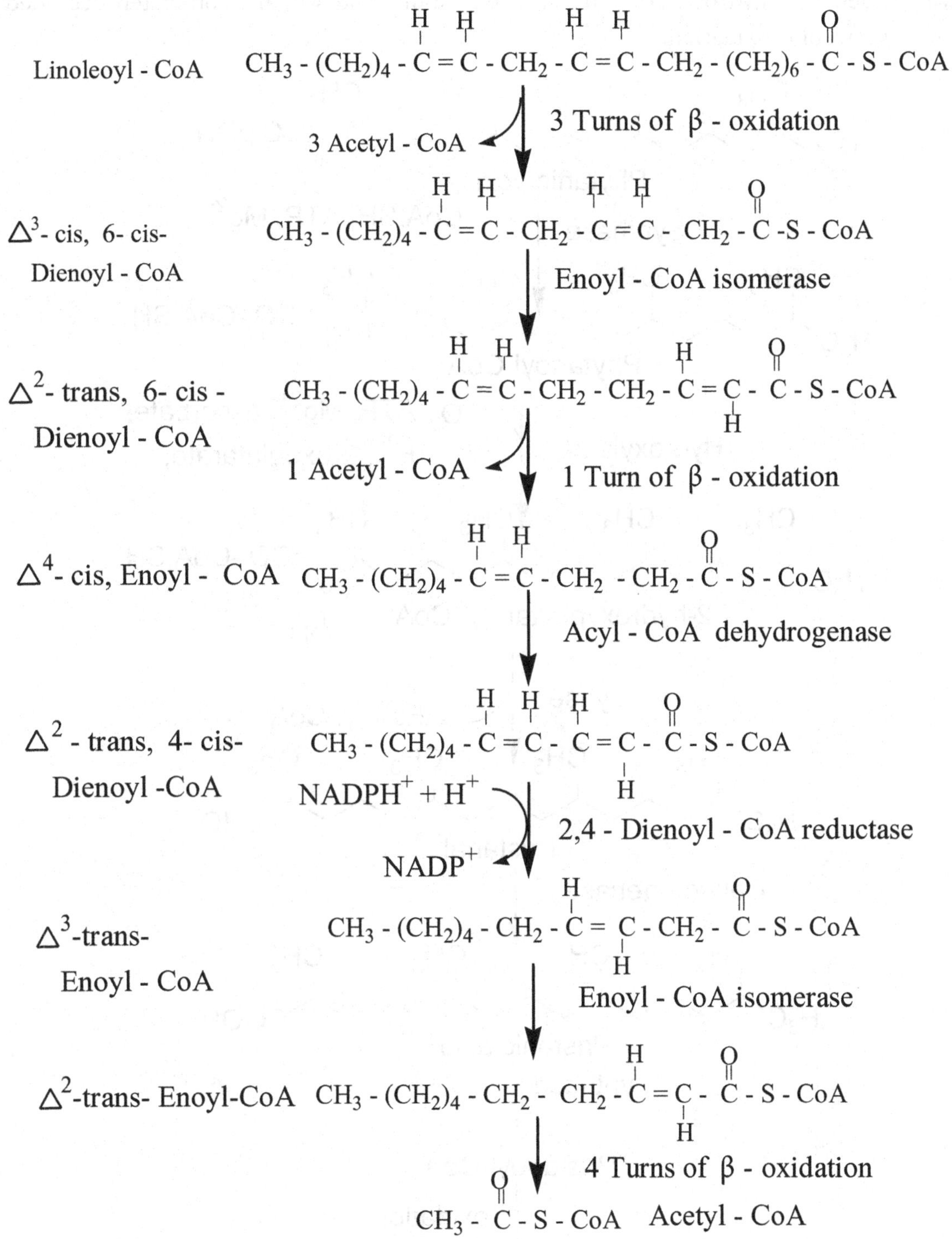

Linoleoyl - CoA $CH_3 - (CH_2)_4 - \overset{H}{C} = \overset{H}{C} - CH_2 - \overset{H}{C} = \overset{H}{C} - CH_2 - (CH_2)_6 - \overset{O}{C} - S - CoA$

3 Acetyl - CoA ← 3 Turns of β - oxidation

Δ^3- cis, 6- cis-
Dienoyl - CoA $CH_3 - (CH_2)_4 - \overset{H}{C} = \overset{H}{C} - CH_2 - \overset{H}{C} = \overset{H}{C} - CH_2 - \overset{O}{C} - S - CoA$

Enoyl - CoA isomerase

Δ^2- trans, 6- cis -
Dienoyl - CoA $CH_3 - (CH_2)_4 - \overset{H}{C} = \overset{H}{C} - CH_2 - CH_2 - \overset{H}{C} = \overset{H}{\underset{H}{C}} - \overset{O}{C} - S - CoA$

1 Acetyl - CoA ← 1 Turn of β - oxidation

Δ^4- cis, Enoyl - CoA $CH_3 - (CH_2)_4 - \overset{H}{C} = \overset{H}{C} - CH_2 - CH_2 - \overset{O}{C} - S - CoA$

Acyl - CoA dehydrogenase

Δ^2 - trans, 4- cis-
Dienoyl -CoA $CH_3 - (CH_2)_4 - \overset{H}{C} = \overset{H}{C} - \overset{H}{C} = \overset{H}{C} - \overset{O}{C} - S - CoA$

NADPH$^+$ + H$^+$ ⤵ 2,4 - Dienoyl - CoA reductase

NADP$^+$ ←

Δ^3-trans-
Enoyl - CoA $CH_3 - (CH_2)_4 - CH_2 - \overset{H}{C} = \overset{H}{\underset{H}{C}} - CH_2 - \overset{O}{C} - S - CoA$

Enoyl - CoA isomerase

Δ^2-trans- Enoyl-CoA $CH_3 - (CH_2)_4 - CH_2 - CH_2 - \overset{H}{C} = \overset{H}{\underset{H}{C}} - \overset{O}{C} - S - CoA$

4 Turns of β - oxidation

$CH_3 - \overset{O}{C} - S - CoA$ Acetyl - CoA

Figure 4.8

The oxidation of phytanic acid in peroxisomes involves activation, α-hydroxylation, cleavage, and aldehyde dehydrogenation to generate pristanic acid, which is completely degraded by three turns of ß-oxidation.

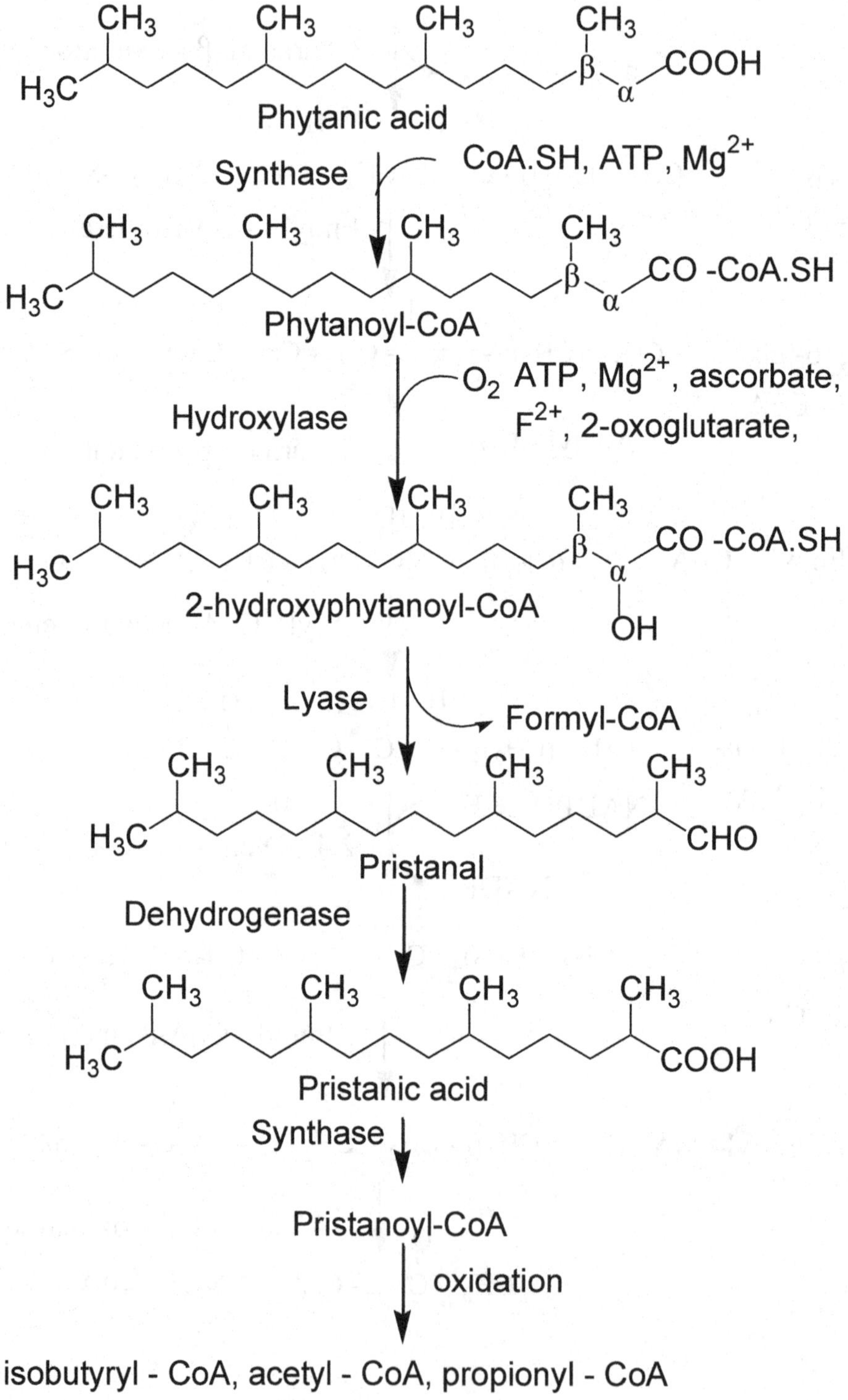

Figure 4.9

Oxidation of ω–fatty acids. The ω-carbon of the fatty acid primarily undergoes hydroxylation by CYP4A11 and is then oxidized by aldehyde and alcohol dehydrogenase. The dicarboxylic acid product is then completely oxidized by ß-oxidation.

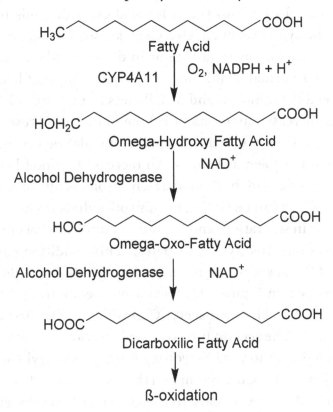

Figure 4.10

The ackee fruit. Used with permission from Doc Taxon, 2013.

KETONE-BODY SYNTHESIS (KETOGENESIS)

The acetyl-CoA formed in liver mitochondria from fatty acids can undergo a number of metabolic fates. One is the formation of ketone bodies (acetoacetate, ß-hydroxybutyrate, and acetone) when fat and carbohydrate degradation are not appropriately balanced. Entry of acetyl-CoA into the citric-acid cycle depends on the availability of oxaloacetate for the synthesis of citrate. Thus, ketone bodies are formed when the levels of oxaloacetate are low, as in fasting or in diabetes, when oxaloacetate is used to form glucose. Under normal conditions, the concentration of ketone bodies in the blood does not exceed 0.2 mmol/L, and usually less than 1 mg is lost in the urine in 24 hours, but when increased amounts of ketone bodies are present in the blood or urine, the body experiences ketonemia (hyperketonemia) or ketonuria, respectively. The overall condition is known as ketosis. An increase in blood levels of acetoacetic acid and ß-hydroxybutyric acid, both moderately strong acids, may lower the pH and lead to ketoacidosis, which can be fatal in uncontrolled diabetes mellitus. The simplest form of ketosis occurs in starvation, and the most complicated occurs in patients with uncontrolled diabetes mellitus (type-1 diabetes). This condition can be harmful because of the effect of the lower pH on binding of oxygen to hemoglobin.

Two pathways (shown in figure 4.11) form acetoacetate from 2 molecules of acetyl-CoA. These are catalyzed by acetoacetyl-CoA thiolase. The first is simple deacylation of acetoacetyl-CoA. The second involves condensation of acetoacetyl-CoA with a third molecule of acetyl-CoA to yield ß-hydroxy-ß-methylglutaryl-CoA (HMG-CoA) in a reaction catalyzed by HMG-CoA synthase. HMG-CoA lyase then causes the release of acetyl-CoA from HMG-CoA to leave free acetoacetate. Acetoacetate is reduced to ß-hydroxybutyrate by ß-hydroxybuyrate dehydrogenase. Acetoacetate spontaneously decarboxylates to form acetone, but in a very small amounts.

Use of ketone bodies

Ketone bodies are used in extrahepatic tissues because the enzymes required for their activation are absent in the liver – they are released from liver mitochondria and are conveyed to the cytosol of extrahepatic tissues without need for a carrier because they are soluble in aqueous solution. Acetoacetate and ß-hydroxybutyrate are converted to acetyl-CoA (figure 4.12).

Figure 4.11

The pathways of ketogenesis.

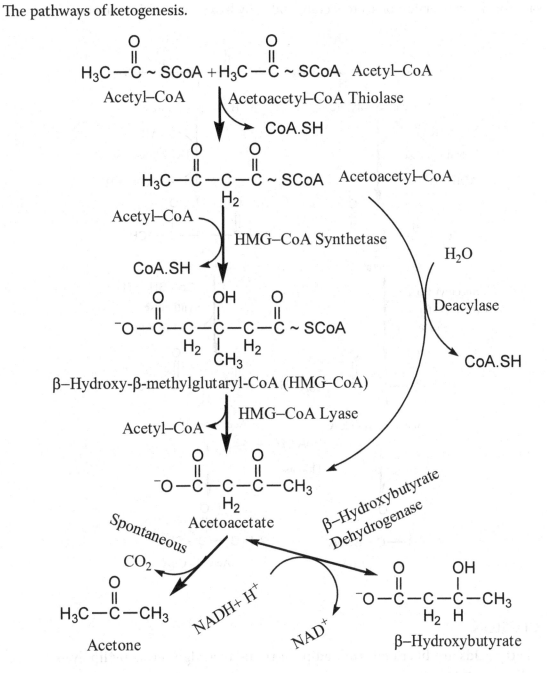

Figure 4.12

Pathways for the conversion of acetoacetate and ß-hydroxybutyrate into acetyl-CoA.

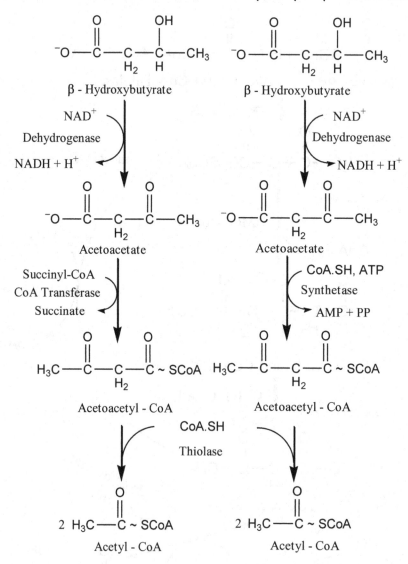

CONCLUSION

- Fatty acids are liberated from adipose-tissue triacylglycerols by lipolysis before they are oxidized.
- Catecholamines, glucagon, parathyroid hormone, thyrotropin, α-melano-cyte-stimulating hormone, and adrenocorticotropin enhance triacylglycerol degradation, whereas insulin decreases it.
- Fatty acids enter into the mitochondrial matrix by a carnitine shuttle to undergo ß-oxidation. Minor pathways for their oxidation are α-oxidation (for phytanic acid), microsomal ω-oxidation (for 10 or 12 carbon acids), and peroxisomal oxidation (for phytanic and very-long-chain fatty acids).

- Disorders that arise from impaired fatty acid oxidation include carnitine-palmitoyl-CoA transferase I deficiency, Jamaican vomiting sickness, Zellweger's syndrome, dicarboxylic aciduria, X-linked adrenoleukodystrophy, and Refsum's disease.
- Ketone bodies (acetoacetate, ß-hydroxybutyrate, and acetone) are formed during prolonged fasting or in diabetes mellitus and lead to ketosis, ketonemia, and ketonuria.

SELECTED READINGS

Casteels, M., Foulon, V., Mannaerts, G. P., et al. (2003). Alpha-oxidation of 3-methyl-substituted fatty acids and its thiamine dependence. *European Journal Biochemistry* 270 (8): 1619–1627.

Eaton, S., Bartlett, K. and Pourfarzam, M. (1996). Mammalian mitochondrial ß-oxidation. *Biochemical Journal* 320: 345–357.

Goodridge, G. (1991). Fatty acid synthesis in eucaryotes. In D. E. Vance and J. E. Vance (eds.), *Biochemistry of lipids, lipoproteins, and membranes.* Amsterdam: Elsevier Science Publishers, 111–139.

Grunes, D. E., Scordi-Bello, I., Suh, M., et al. (2012). Fulminant hepatic failure attributed to ackee fruit ingestion in a patient with sickle cell trait. *Case Reports Transplantation* doi:10.1155/2012/739238.

Hammarström, S. (1983). Le ukotrienes. *Annual Review of Biochemistry,* 52: 355–377.

Jethi, R. K., and Singh, R. (1976). Bioenergetics of ß-oxidations of fatty acids. *Biochemical Education.* 4 (3): 51.

Moncada, S. (ed.). (1983). Prostacyclin, thromboxanes, and leukotrienes. *British Medical Bulletin* 39: 209–295.

Moser, H. W., Smith, K. D., Moser, A. B. (1995). X-Linked adrenoleukodystrophy. In C. R. Scriver, A. L. Beaudet, W. S. Sly, et al. (eds.). *The metabolic and molecular basis of inherited disease,* 7th ed., vol. 2. New York: McGraw-Hill, 2325–2424.

Perera, N. J., Lewis, B., Tran, H., et al. (2011). Case report Refsum's disease: Use of the intestinal lipase inhibitor Orlistat as a novel therapeutic approach to a complex disorder. *Journal of Obesity* doi: 10.1155/2011/482021.

Ronald, J. A., Wanders, R. J. A., Komen, J., et al. (2010). Fatty acid omega-oxidation as a rescue pathway for fatty acid oxidation disorders in humans. *FEBS Journal* 278 (2011): 182–194.

Saini-Chohan, H. K., Mitchell, R. W. (2012). Delineating the role of alterations in lipid metabolism to the pathogenesis of inherited skeletal and cardiac muscle disorders. *Journal of Lipid Research* 53(1): 4–27.

Schulz, H. (1985). Oxidation of fatty acids. In D. E. Vance and J. E. Vance (eds.), *Biochemistry of lipids and membranes.* Menlo Park, CA: Benjamin/Cummings, 116–142.

Smith, W. L., and Borgeat, P. (1985). The eicosanoids: prostaglandins, thromboxanes, leukotrienes, and hydroxyeico-saenoic acids. In D. E. Vance and J. E. Vance (eds.), *Biochemistry of lipids and membranes.* Menlo Park, CA: Benjamin/Cummings, 325–360

Van Veldhoven, P. P. (2010). Biochemistry and genetics of inherited disorders of peroxisomal fatty acid metabolism. *Journal of Lipid Research* 51 (10): 2863–2895.

Wanders, S. M., and Houten, R. J. (2010). A general introduction to the biochemistry of mitochondrial fatty acid β-oxidation. *Journal of Inherited Metabolic Disease* 33: 469–477.

Zchner, R., Kienesberger, P. C., Haemmerle, G., et al. (2009). Adipose triglyceride lipase and the lipolytic catabolism of cellular fat stores. *The Journal of Lipid Research* 50: 3–21.

Chapter 5: Metabolism of Acylglycerols and Sphingolipids

Introduction

This chapter deals with biosynthesis and degradation of acylglycerols and sphingolipids. Since degradation of triacylglycerols (TAG) has already been described, only biosynthesis of TAG will be discussed here. TAG constitutes most lipids in the human body, particularly adipose tissue. Acylglycerols, especially phospholipids, are major constituents of cell membranes. We will see that phospholipids are involved in metabolism of several other lipids. Sphingolipids, present primarily in the nervous tissue, account for less than 10% of the total lipid content of cell membranes. Sphingolipids, glycosphingolipids, and sphingomyelins are defined by their 18-carbon amino-alcohol backbones. Sphingolipids play significant roles in membrane biology and provide many bioactive metabolites that regulate cell function. Ceramide, sphingosine, and sphingosine-1-phosphate are known to act as bioactive signaling molecules during cell growth, differentiation, senescence, and apoptosis. Despite the diversity of structures and functions of sphingolipids, their creation and destruction are governed by common synthetic and catabolic pathways.

Metabolism of Acylglycerols

Glycerol and fatty acids need to be activated by glycerol kinase and acyl-CoA synthetase, respectively, before they can be incorporated into acylglycerols. Where glycerol kinase is not present or has low activity, as in muscle and adipose tissue, most glycerol 3-phosphate is supplied by dihydroxyacetone phosphate (an intermediate of glycolysis). Glycerol 3-phosphate dehydrogenase catalyzes the conversion of dihydroxyacetone phosphate into glycerol 3-phosphate (figure 5.1).

Biosynthesis of Phosphatidate

Phosphatidate (diacylglycerol 3-phosphate) is the precursor to the biosynthesis of phosphoglycerides and TAG. Phosphatidate is synthesized when 2 molecules of acyl-CoA are combined with 1 molecule of glycerol 3-phosphate. This occurs in two steps (figure 5.2), the first catalyzed by glycerol-3-phosphate acyltransferase to produce lysophosphatidate, and the second catalyzed by 1-acylglycerol 3-phosphate acyltransferase (lysophosphatidate acyltransferase) to yield phosphatidate.

BIOSYNTHESIS OF TRIACYLGLYCEROLS (TAG)

As noted previously, phosphatidate serves as a precursor for both glycerophospholipids and TAG. In the pathway for the synthesis of TAG, phosphatidate is hydrolyzed by phosphatidate phosphohydrolase to yield 1,2-diacylglycerol. This intermediate is finally acylated to produce TAG in a reaction catalyzed by diacylglycerol acyltransferase (figure 5.3). These two enzymes are known as TAG synthetase complex, which is normally bound to the endoplasmic reticulum of the cell. TAG is synthesized in the intestinal mucosa by another pathway, the 2-monoacylglycerol pathway, which takes place in the small intestine by two successive acylations.

BIOSYNTHESIS OF GLYCEROPHOSPHOLIPIDS

Synthesis of glycerophospholipids (or phosphoglycerides) proceeds along several routes. As noted before, phosphatidic acid is the simplest glycerophospholipid and is the precursor for other members of this group. Most eukaryotic cells contain six classes of glycerophospholipids. These classes are phosphatidylethanolamine (cephalin), phosphatidylcholine (lecithin), phosphatidylserine, phosphatidylinositol, phosphatidylglycerol, and cardiolipin.

Figure 5.1

Metabolism of glycerol 3-phosphate.

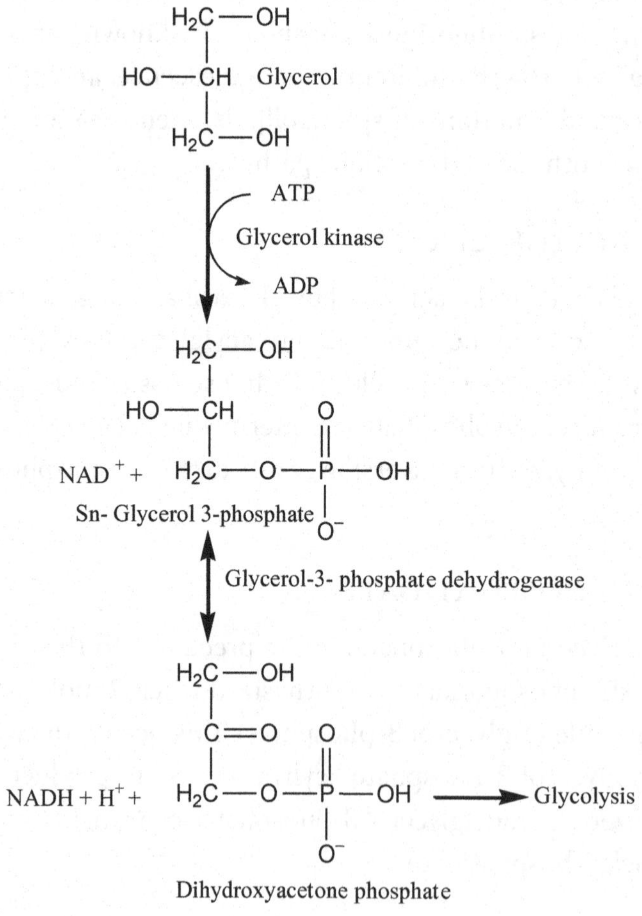

Biosynthesis of phosphatidylcholine (PC) and phosphatidylethanolamine (PE)

The pathways leading to biosynthesis of these glycerophospholipids are shown in figures 5.4 and 5.5. Both can be synthesized by pathways that use free choline or ethanolamine, respectively, which must first be phosphorylated by ATP to phosphorylcholine and phosphorylethanolamine, respectively. They then react with cytidine triphosphate (CTP) to form cytidine diphosphocholine (CDP-choline) and cytidine diphosphoethanolamine (CDP-ethanolamine), respectively. Finally, a phosphorylated base (phosphorylcholine or phosphorylethanolamine) is transferred to diacylglycerol to form phosphatidylcholine and phosphatidylethanolamine, respectively.

Figure 5.2

Biosynthesis of phosphatidate.

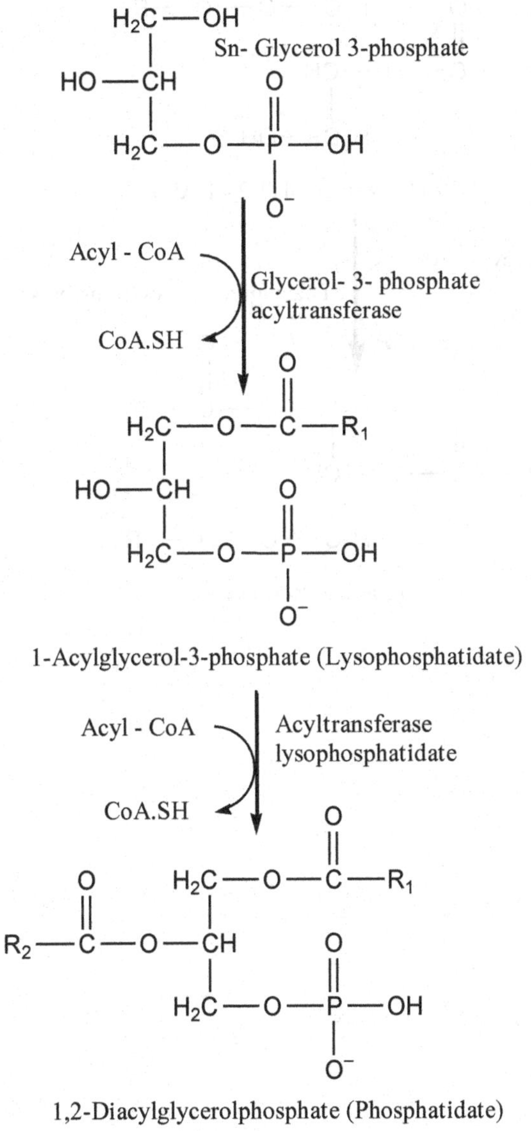

1-Acylglycerol-3-phosphate (Lysophosphatidate)

1,2-Diacylglycerolphosphate (Phosphatidate)

Figure 5.3

Biosynthesis of triacylglycerol (TAG).

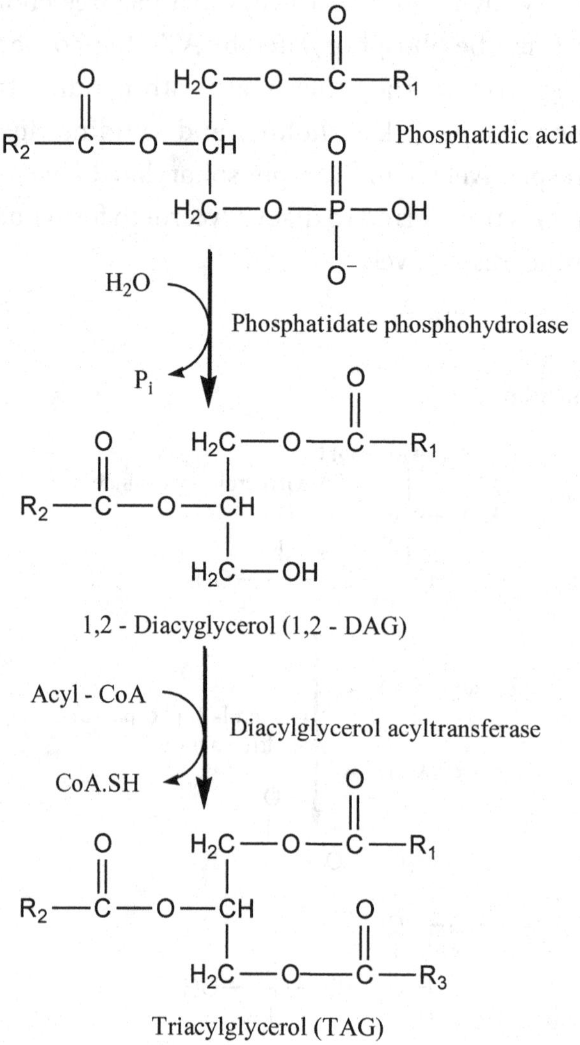

Triacylglycerol (TAG)

Figure 5.4

Biosynthesis of phosphatidylcholine (PC).

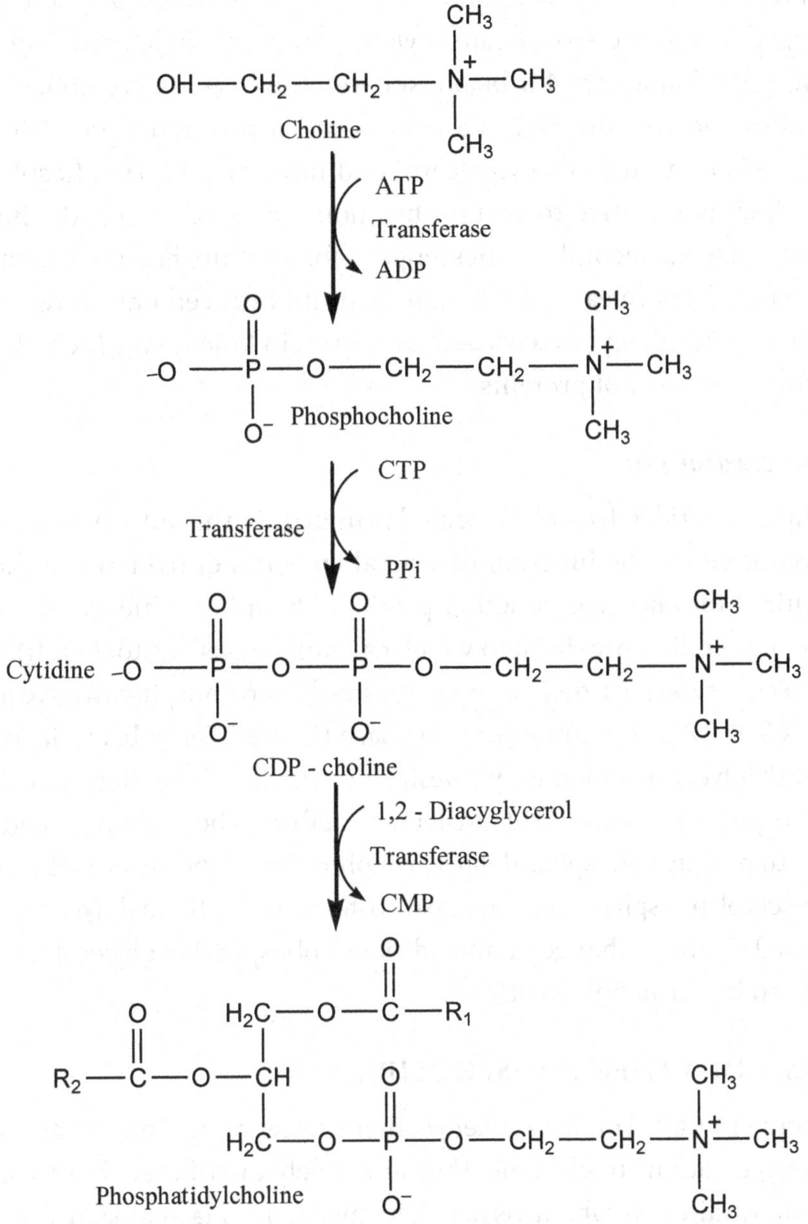

Both phosphatidylcholine and phosphatidylethanolamine are synthesized from phosphatidylserine (see figure 5.6) after it is converted to phosphatidylethanolamine by phosphatidylserine decarboxylase. Phosphatidylethanolamine-serine transferase exchanges a free ethanolamine for the serine moiety of phosphatidylserine to produce serine and phosphatidylethanolamine. This reaction is reversible, so phosphatidylserine can also be formed from phosphatidylethanolamine. In the liver, phosphatidylethanolamine is converted into phosphatidylcholine by 3 successive methylation reactions by the same enzyme, which uses S-adenosylmethionine as the methyl-group donor (figure 5.6).

Biosynthesis of phosphoinositides

This synthesis starts when CDP-diacylglycerol reacts by action of phosphatidylinositol synthase with L-myo-inositol to form phosphatidylinositol (figure 5.7). The phosphatidylinositol undergoes two successive phosphorylations to form phosphatidylinositol 4-phosphate and phosphatidylinositol 4,5-bisphosphate. These lipids are collectively termed phosphoinositides and are enriched in arachidonate at position 2. Phosphatidylinositol 4,5-bisphosphate is cleaved to diacylglycerol and inositol 1,4,5-trisphosphate by phospholipase C, which is sensitive to certain hormones and other signals. Both products play important roles as secondary messengers for certain hormones. Inositol-1,4,5-trisphosphate stimulates release of calcium from its intracellular stores (endoplasmic reticulum), whereas 1,2-diacylglycerol activates protein kinase C, which phosphorylates an extremely diverse group of proteins.

Biosynthesis of cardiolipin

Cardiolipin (diphosphatidylglycerol) is located primarily in the mitochondrial inner membrane and is required for the function of several enzymes of oxidative phosphorylation. Thus, it is required for energy production particularly in heart muscle. Cardiolipin also serves as a calcium binding site through which calcium triggers mitochondrial-membrane permeability. In eukaryotes, cardiolipin is synthesized from phosphatidylglycerol and CDP-diacylglycerol (CDP-DAG) by cardiolipin synthase (figure 5.8). In bacteria, it is produced from phosphatidylglycerol, which is synthesized from CDP-DAG (figure 5.9). In the first step, glycerophosphate phosphatidyl transferase catalyzes the exchange of glycerol phosphate for CMP to produce phosphatidylglycerol phosphate. The latter is then catalyzed by phosphatidylglycerol phosphate phosphatase to form phosphatidylglycerol. Finally, cardiolipin is formed by the exchange of an additional phosphatidylglycerol for glycerol in a reaction catalyzed by cardiolipin synthase.

BIOSYNTHESIS OF ETHER PHOSPHOLIPIDS

These contain either O-alkyl or an O-alkenyl ether species rather than an acyl group linked to one of the oxygen atoms of glycerol. They are widely distributed, but their abundance varies greatly depending on which tissue they appear in. Plasmalogens, or vinyl ethers that contain alkenyl ether at carbon 1 of the glycerol moiety, constitute nearly 50% of all phospholipids in heart tissue.

Figure 5.10 shows the biosynthetic pathway for ether phospholipids. The precursor of the glycerol moiety is dihydroxyacetone phosphate, which combines with acyl-CoA to produce 1-acyldihydroxyacetone phosphate by the action of acyltransferase. An exchange reaction then replaces the 1-acyl group with an alkyl group derived from an alcohol to form 1-alkyl-dihydroxyacetone phosphate. In presence of NADPH, this is converted to 1-alkylglycerol-3-phosphate, which is then acylated in the 2 position to form the

1-alkyl-2-acylglycerol 3-phosphate (analogous to phosphatidate). This, in turn, is hydrolyzed to 1-alkyl-2-acylglycerol. CDP-ethanolamine can react with 1-alkyl-2-acylglycerol to yield 1-alkyl-2-acylglycerol 3-phosphethanolamine, which is converted to 1-alkenyl 2-acylglycerol 3-phosphethanolamine (plasmalogen) by a desaturase in the presence of O_2, NADH, and cytochrome b_5.

Figure 5.5

Biosynthesis of phosphatidylethanolamine (PE).

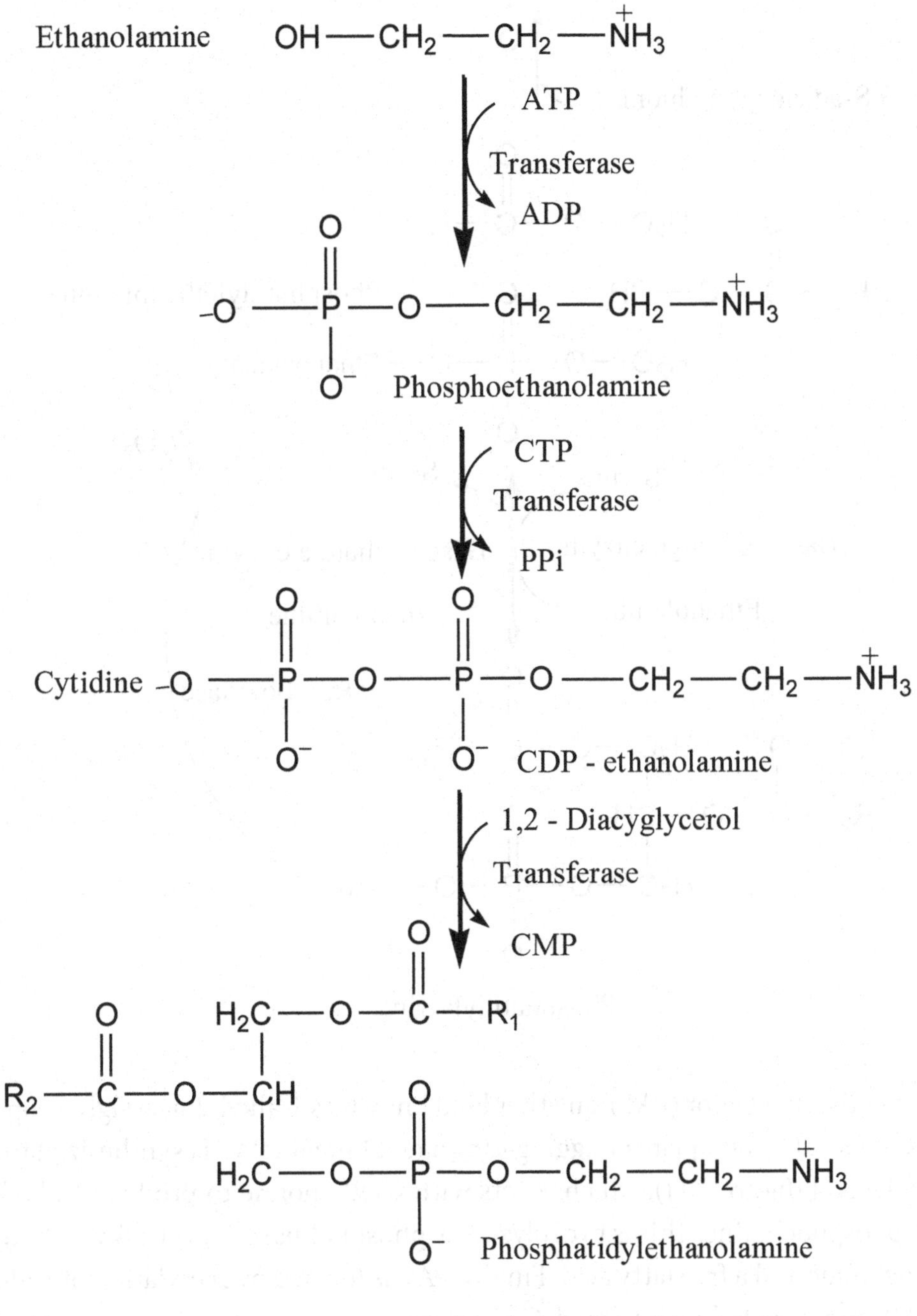

69

Figure 5.6

Interconversion of glycerophospholipids

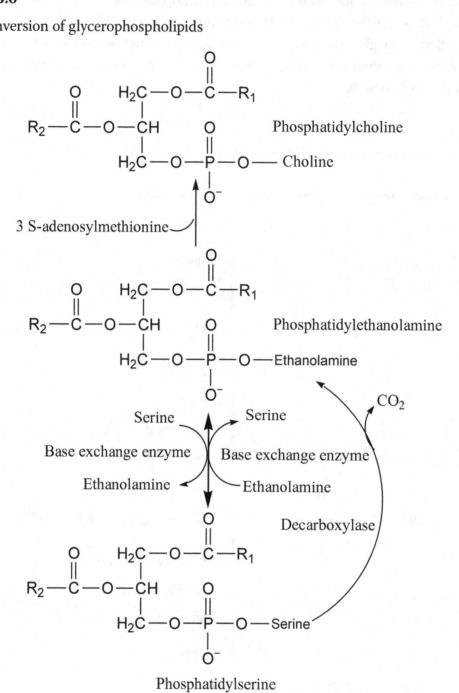

Phosphatidylserine

Platelet-activating factor (PAF), an ether lipid known as 1-alkyl-2-acetylglycerophosphocholine, is the most potent platelet-aggregating agent known. PAF is synthesized from 1-alkyl-2-acylglycerol (figure 5.11), which reacts with CDP-choline to produce 1-alkyl-2-acylglycerol 3-phosphocholine. This is hydrolyzed by phospholipase A_2 to 1-alkyl-2-lysoglycerol 3-phosphocholine and a free fatty acid. Finally, PAF is formed by acetylation of 1-alkyl-2-lysoglycerol 3-phosphocholine by acetyl-CoA transferase.

DEGRADATION OF GLYCEROPHOSPHOLIPIDS

Glycerophospholipids are degraded by phospholipases that hydrolyze specific bonds. These enzymes are present in all types of cells and in various subcellular locations within eukaryotic cells. They are classified according to which bond is cleaved in a glycerophospholipid molecule (see figure 5.12). Phospholipases A_1 and A_2 selectively remove fatty acids from the sn-1 and sn-2 positions, respectively. Phospholipase C releases the phosphorylated nitrogen moiety from the glycerophospholipid molecules. Phospholipase D, which occurs mainly in plants, releases the nitrogen base from phospholipids. Phospholipase A_2 is a major constituent of some snake venoms that rupture red blood cells, and can cause hemolysis.

When phosphatidylcholine is attacked by phospholipase A_2, it releases a free fatty acid and lysolecithin (lysophosphatidylcholine). An acyltransferase can resynthesize phosphatidylcholine from lysolecithin. Alternatively, lysolecithin can be hydrolyzed by phospholipase A_1 to free fatty acid and glycerophosphocholine, which can be hydrolyzed to glycerol-3-phosphate and choline. Figure 5.13 shows the metabolism of phosphatidylcholine. Lysolecithin can also be formed by lecithin-cholesterol acyltransferase (LCAT), which transfers a fatty acid residue from the 2 position of lecithin to cholesterol to yield cholesterol ester and lysolecithin.

$$\text{Lecithin + Cholesterol} \xrightarrow{\text{LCAT}} \text{Cholesterol ester + Lysolecithin}$$

LCAT is produce in the liver and released in plasma, which is responsible for the formation of all plasma-cholesterol ester in humans. Note that lysolecithin in plasma is a good detergent and is capable of solubilizing cell membranes, particularly those of erythrocytes, causing them to lyse.

BIOSYNTHESIS OF SPHINGOLIPIDS

All sphingolipids are formed from ceramide, which can be produced in at least two ways: first through the de novo pathway, and second through hydrolysis of complex lipids, especially sphingomyelin (SM).

De novo synthesis of ceramide starts with the reaction of acyl-CoA and sphingosine (the backbone of all sphingolipids). Sphingosine synthesis occurs in the endoplasmic reticulum and starts with the condensation of palmitoyl-CoA and serine by serine palmitoyltransferase in the presence of pyridoxal-5-phosphate (PLP) as coenzyme to form 3-ketodihydrosphingosine. In the presence of NADPH, this is then reduced by 3-ketodihydrosphingosine reductase to dihydrosphingosine, which is oxidized to sphingosine by a flavoprotein enzyme. Ceramide is formed by acylation of the amino group of sphingosine (see figure 5.14).

Figure 5.7

Biosynthesis of phosphoinositides (PI).

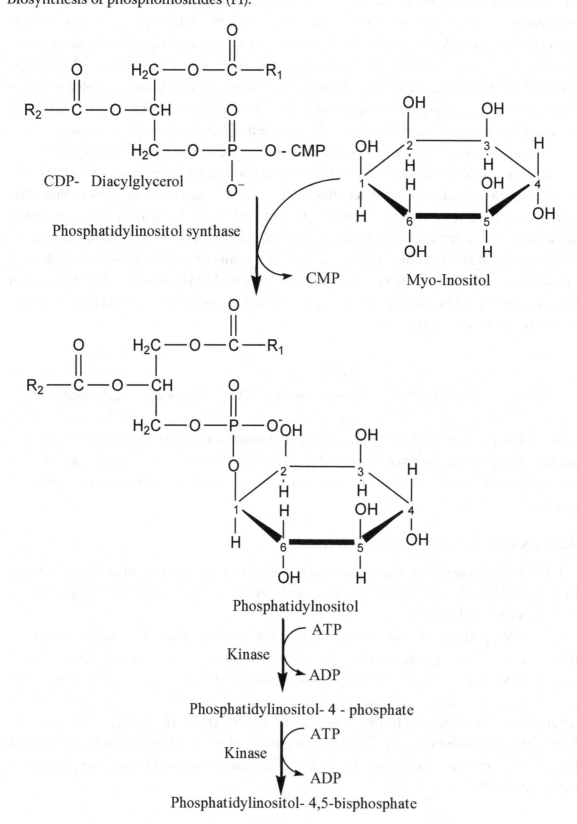

Figure 5.8

Biosynthesis of cardiolipin in eukaryotes.

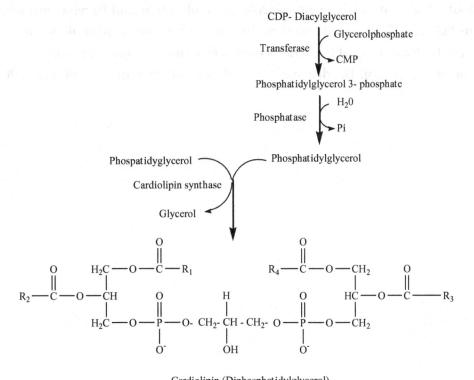

Figure 5.9

Biosynthesis of cardiolipin in prokaryotes.

Sphingomyelins are the most abundant complex sphingolipids in human cells. They are generated when ceramide reacts with either CDP-choline or phosphatidylcholine by action of sphingomyelin synthases (see figure 5.15). The sphingomyelin of the myelin sheet contains mainly long-chain fatty acids, such as lignoceric, and nervonic acids, whereas in the grey matter of the brain, sphingomyelin contains primarily stearic acid.

Degradation of sphingomyelin occurs through hydrolysis of the phosphocholine head groups by a sphingomyelinase with production of ceramide and phosphocholine.

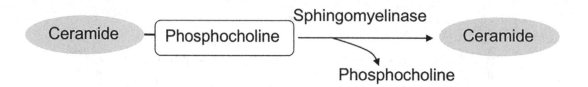

Niemann-Pick disease (NPD) refers to a group of severe inherited metabolic disorders resulting in lysosomal accumulation of sphingomelin, cholesterol, and other metabolically related lipids. Types A and B of Niemann-Pick disease are two clinically distinct forms of lysosomal storage disorders that result from the inherited deficiency of acid sphingomyelinase activity, with the consiequence being sphingomyelin accumulation in lysosomes. The two types are described according to the presence (type A) or absence (type B) of neurologic symptoms. Type A is fatal disorder of infancy characterized by failure to thrive, hepatosplenomegaly, and rapidly progressive neurologic deterioration. In contrast, most type-B-NPD individuals have little or no neurologic involvement and survive into adulthood. Niemann-Pick type C is a fatal autosomal-recessive lipidosis distinguished by a unique error in cellular trafficking of exogenous cholesterol that is associated with lysosomal accumulation of cholesterol. The disease is caused by a mutation in the NPC1 or NPC2 gene.

Figure 5.10

Biosynthesis of ether phospholipids.

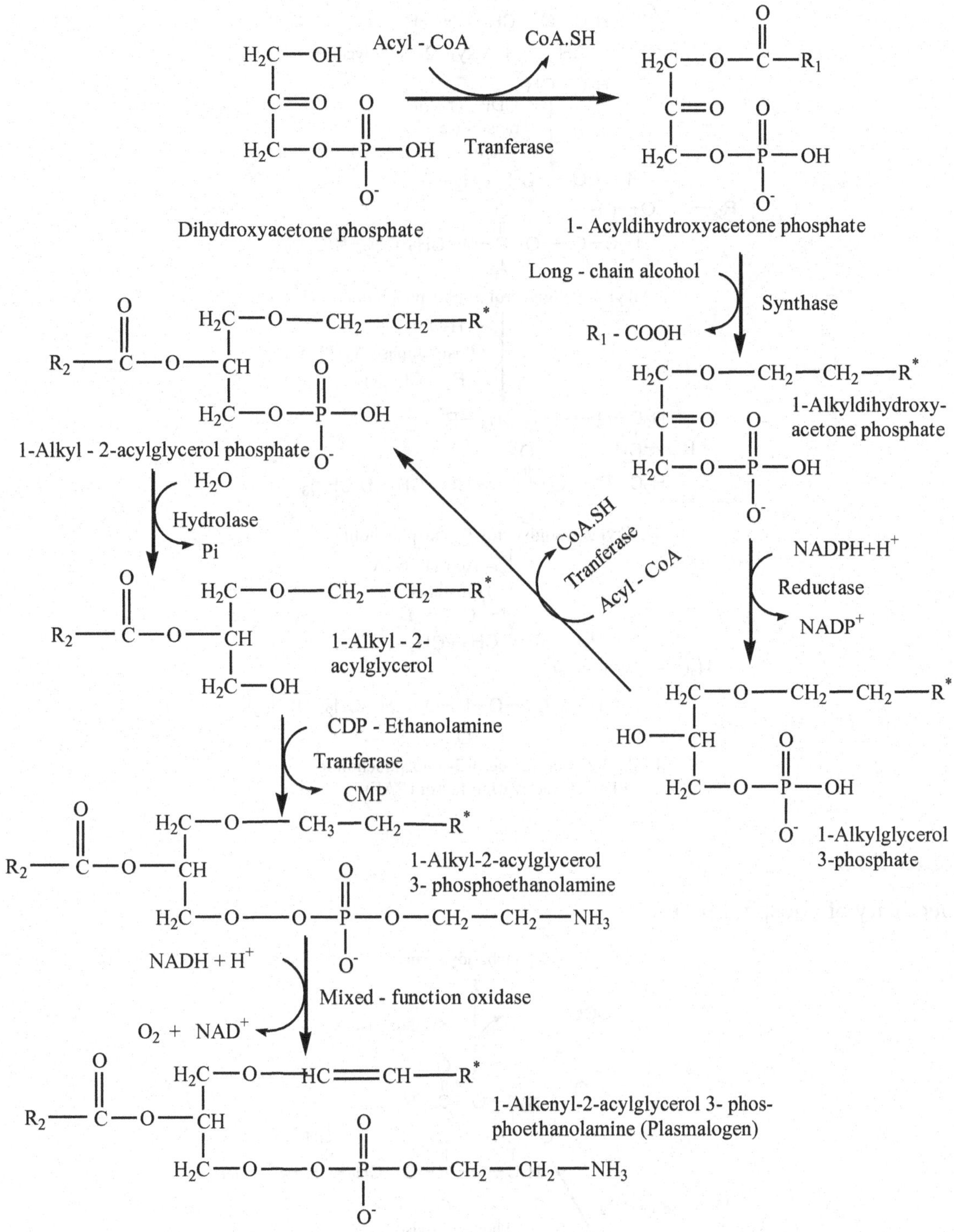

Figure 5.11

Biosynthesis of platelet-activating factor (PAF).

1-Alkyl - 2- acylglycerol

CDP - choline Transferase
CMP

1-Alkyl-2-acylglycerol 3- phosphocholine

H_2O
Phospholipase A_2 (PLA$_2$)
R_2—COOH

1-Alkyl-2-lysoglycerol 3- phosphocholine

Acetyl - CoA Transferase
CoA.SH

1-Alkyl-2-acetylglycerol 3- phosphocholine
Platelet - activating factor (PAF)

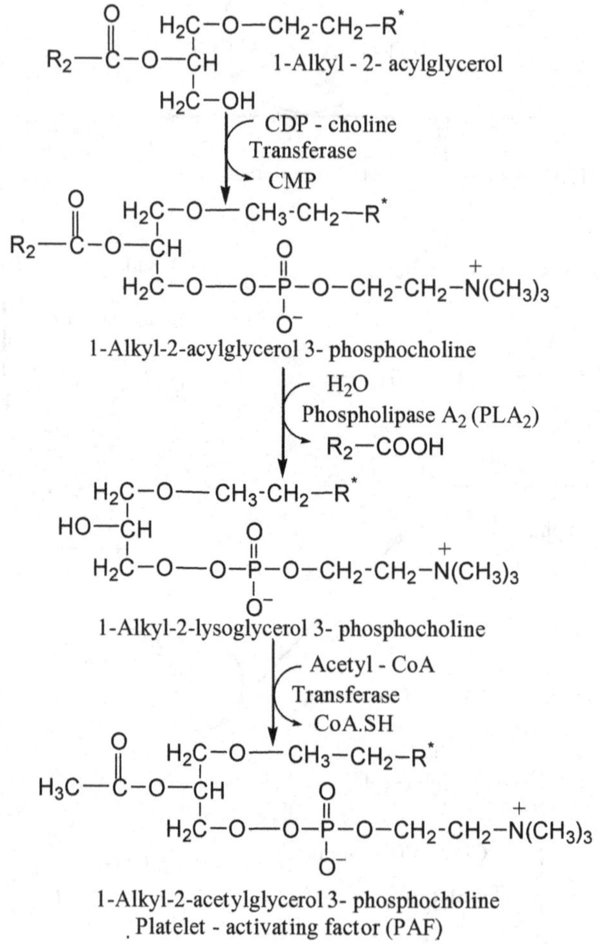

Figure 5.12

Specificity of phospholipases.

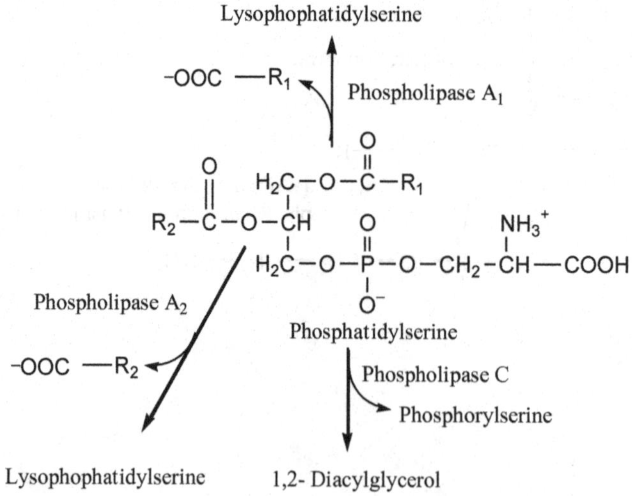

Lysophophatidylserine
-OOC —R$_1$
Phospholipase A$_1$

Phospholipase A$_2$
-OOC —R$_2$

Phosphatidylserine

Phospholipase C
Phosphorylserine

Lysophophatidylserine   1,2- Diacylglycerol

Figure 5.13

Metabolism of phosphatidylcholine.

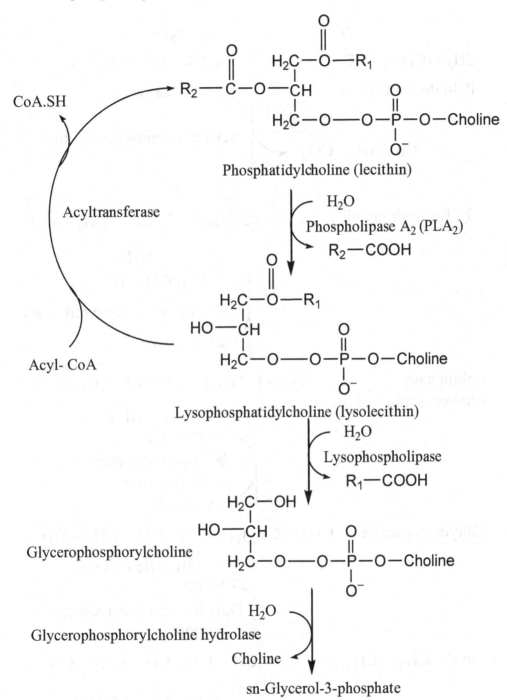

Figure 5.14

Biosynthesis of sphingosine and ceramide.

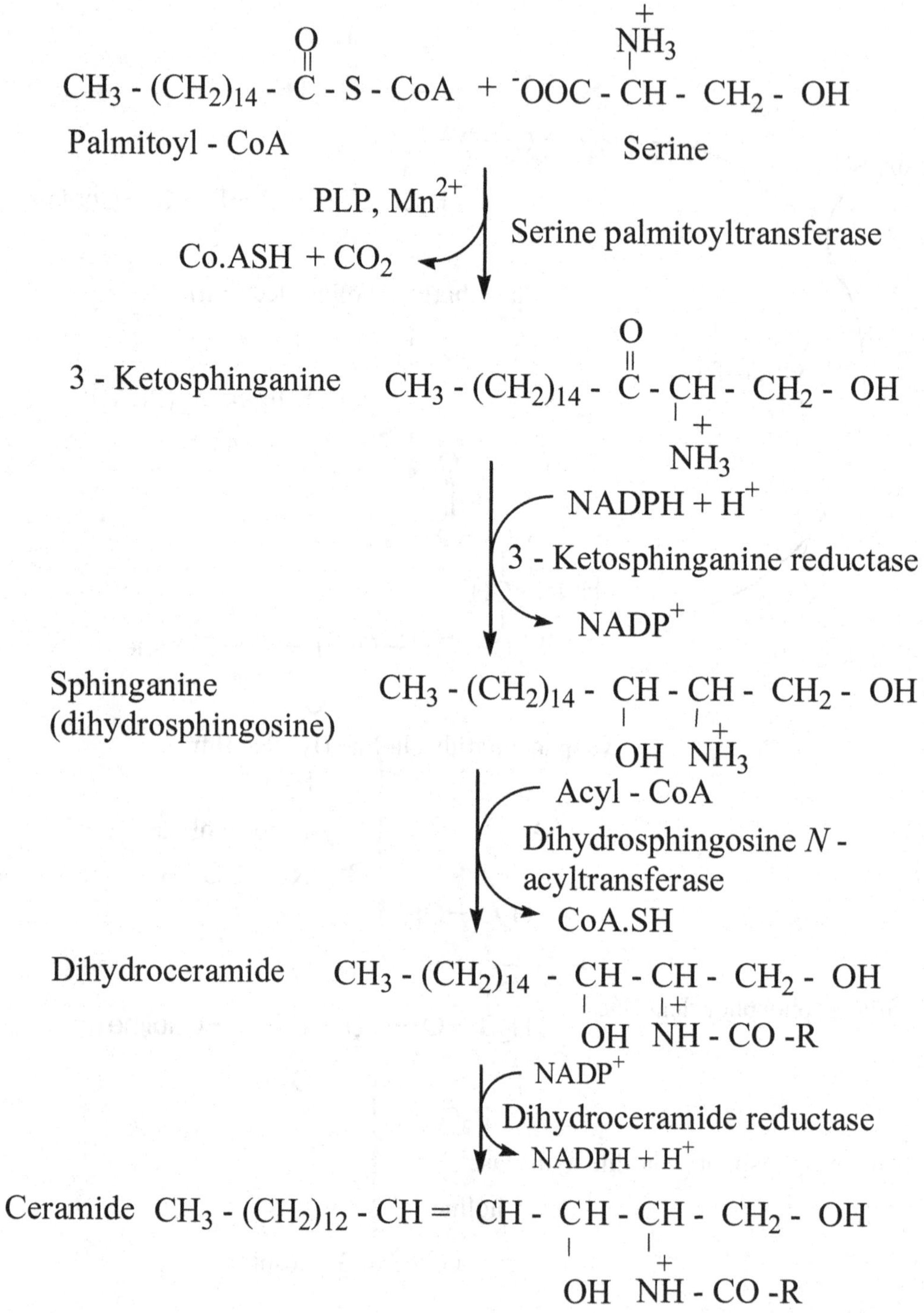

$$CH_3\text{-}(CH_2)_{14}\text{-}\overset{O}{\overset{\|}{C}}\text{-}S\text{-}CoA \;+\; {}^-OOC\text{-}\overset{\overset{+}{NH_3}}{\underset{}{CH}}\text{-}CH_2\text{-}OH$$

Palmitoyl - CoA Serine

PLP, Mn^{2+} — Serine palmitoyltransferase

$Co.ASH + CO_2$

3 - Ketosphinganine $\quad CH_3\text{-}(CH_2)_{14}\text{-}\overset{O}{\overset{\|}{C}}\text{-}\underset{\overset{|}{\overset{+}{NH_3}}}{CH}\text{-}CH_2\text{-}OH$

$NADPH + H^+$

3 - Ketosphinganine reductase

$NADP^+$

Sphinganine
(dihydrosphingosine) $\quad CH_3\text{-}(CH_2)_{14}\text{-}\underset{\overset{|}{OH}}{CH}\text{-}\underset{\overset{|}{\overset{+}{NH_3}}}{CH}\text{-}CH_2\text{-}OH$

Acyl - CoA

Dihydrosphingosine N-acyltransferase

CoA.SH

Dihydroceramide $\quad CH_3\text{-}(CH_2)_{14}\text{-}\underset{\overset{|}{OH}}{CH}\text{-}\underset{\overset{|}{NH\text{-}CO\text{-}R}}{CH}\text{-}CH_2\text{-}OH$

$NADP^+$

Dihydroceramide reductase

$NADPH + H^+$

Ceramide $\quad CH_3\text{-}(CH_2)_{12}\text{-}CH=CH\text{-}\underset{\overset{|}{OH}}{CH}\text{-}\underset{\overset{|}{NH\text{-}CO\text{-}R}}{CH}\text{-}CH_2\text{-}OH$

Figure 5.15

Biosynthesis of sphingomyeline.

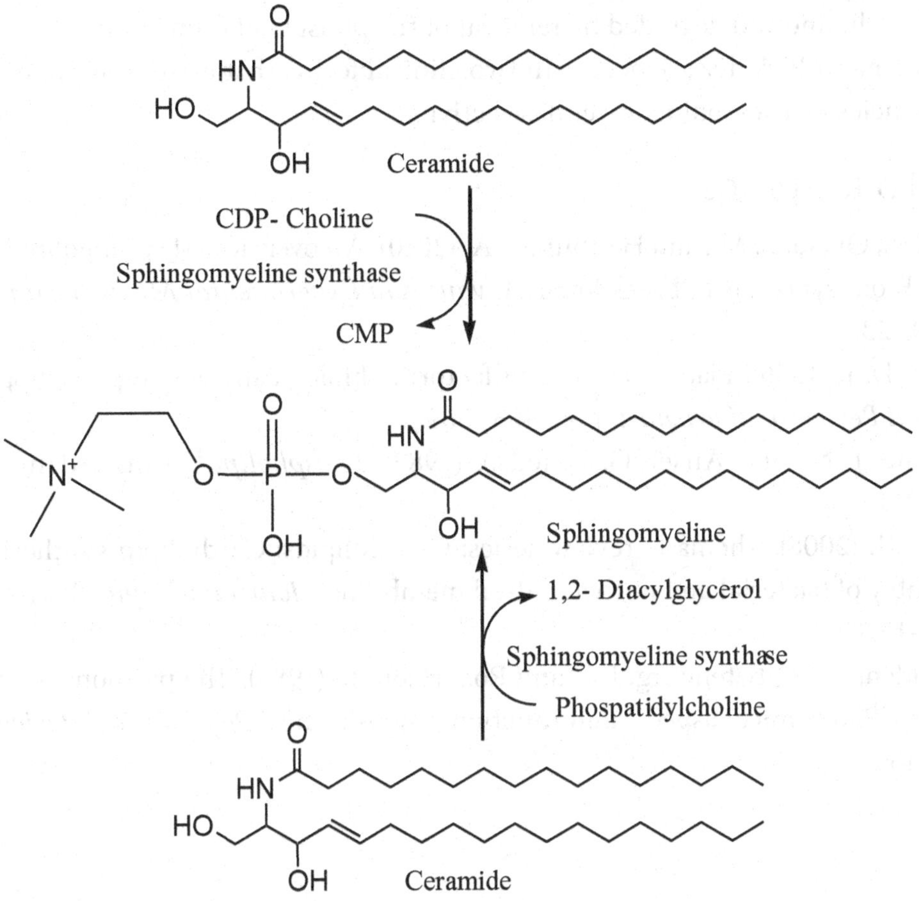

CONCLUSION

- Glycerol and fatty acids are activated before being incorporated into acylglycerols. Where glycerol kinase activity is low, as in muscle and adipose tissue, most glycerol-3-phosphate is derived from dihydroxyacetone phosphate.

- Phosphatidate (diacylglycerol-3-phosphate) is the precursor to the biosynthesis of glycerophospholipids and triacylglycerols.

- Two strategies are used for synthesis of glycerophospholipids: One converts phosphatidic acid to CDP-DAG, which then binds the polar group, whereas the other starts with diacylglycerol and a CDP-polar group, which interact to form the desired product.

- Biosynthesis of phosphoinositides starts when CDP-diacylglycerol reacts with myo-inositol to form phosphatidylinositol and is then phosphorylated to the 4-phosphate and 4,5-bisphosphate forms. When cleaved by phospholipase C, the latter produces the second messengers diacylglycerol and inositol 1,4,5-trisphosphate.

- In degradation of glycerophospholipids, specific phospholipases remove each of the fatty acids and the phosphorylated nitrogen portion of the molecule.
- Sphingomyelins are generated from ceramide and either CDP-choline or phosphatidylcholine and degraded by removal of the phosphocholine group.
- Niemann-Pick disease occurs in two clinical forms that arise from inherited deficiencies of acid-sphingomyelinase activity.

SELECTED READINGS

Gault, C. R., Obeid, L. M., and Hannun, Y. A. (2010). An overview of sphingolipid metabolism: From synthesis to breakdown. *Advances in Experimental Medicune and Biology* 688: 1–23.

Hanahan, D. J. (1986). Platelet activating factor: A biologically active phosphoglyceride. *Annual Review of Biochemistry* 55: 483–509.

Hawthorne, J. N., and Ansel, G. B. (eds.). (1982). *Phospholipid.* Amsterdam: Elsevier, 279–311.

Schlame, M. (2008). Thematic review series: Glycerolipids: Cardiolipin synthesis for the assembly of bacterial and mitochondrial membranes. *Journal of Lipid Research* 49 (8): 1607–1620.

Vane Golden, L. M., Batenburg, J. J., and Robertson, B. (1988). The pulmonary surfactant system: Biochemical aspects and functional significance. *Physiological Reviews* 68 (2): 374–455.

Chapter 6: Metabolism of Glycosphingolipids

Introduction

Glycosphingolipids, a large and heterogeneous family of sphingolipids, are composed of a carbohydrate moiety linked to ceramide. Unlike sphingomyelins, they contain no phosphate, but like sphingomyelin, they are derivatives of ceramide. They are also called glycolipids. The carbohydrate is a monosaccharide, disaccharide, or oligosaccharide which is attached to ceramide through an O-glycosidic linkage. The most important glycosphingolipid classes are the cerebrosides, the sulphatides, and the gangliosides that are essential components of all cell membranes but are present in large quantities in nerve tissue.

Biosynthesis of glycosphingolipids involves the glycosyltransferases, most of which are located on the luminal side of the Golgi apparatus. In contrast, their degradation occurs in lysosomes and is carried out by a family of hydrolytic enzymes. These pathways are of great medical interest because of the congenital diseases called sphingolipidoses (lipid storage diseases), each of which results from deficiency of a degradative enzyme with concomitant accumulation within lysosomes of the substrate for the deficient enzyme and is characterized by severely impaired central-nervous-system function. This chapter discusses the metabolism of some important glycosphingolipids and of some sphingolipidoses.

Biosynthesis of Glycosphingolipids

This process starts from ceramide and takes place in the endoplasmic reticulum and Golgi apparatus. The synthesis of ceramide, which also occurs in the endoplasmic reticulum (ER), was described in chapter 5 (figure 5.15).

Ceramide has very low solubility in an aqueous environment and has to be transported from the ER, where it is synthesized, to the Golgi apparatus, where it can be converted into glycosphingolipids. Ceramide transfer protein (CERT), a cytosolic protein, transfers ceramide from the ER to the Golgi apparatus. An alternative system transports ceramide to the Golgi, is coatomer-protein dependent, is based on vesicular transport.

Glucosylceramide (Glc-Cer) and Galactosylceramide (Gal-Cer) Formation

The stepwise addition of monosaccharide units requires nucleotide-linked sugars as the activated biosynthetic substrates and ceramide as the initial monosaccharide acceptor. The glycosyltransferases are specific to each monosaccharide. Cerebrosides (the simplest glycosphingolipids) are formed when ceramide reacts with UDP-glucose or UDP-galactose to produce glucosylceramide (Glc-Cer) or galactosylceramide (Gal-Cer), respectively. Glucosylceramide synthase attaches glucose to the C1-hydroxyl position of ceramide.

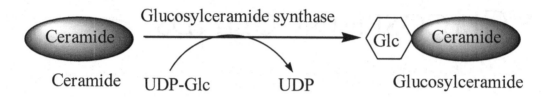

Gal-Cer is an important lipid in myelin. Glc-Cer is the primary glycosphingo-lipid of extraneural tissues and a precursor of the more complex glycosphingolipids. Galactosphingolipids are formed by galactosylceramide transferase.

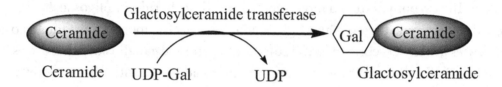

LACTOSYLCERAMIDE PRODUCTION

Most glycolipids in mammals are synthesized from lactosylceramide, which is formed by the addition of a galactose moiety from UDP-Gal to Glc-Cer by galactosylceramide transferase I.

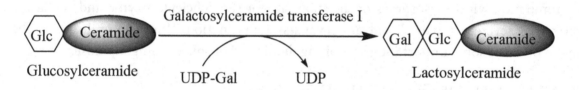

Sulfatides are sulfated Gal-Cer, the sulfate group being transferred from 3-phosphoad-enosine 5'-phosphosulfate (PAPS; "active sulfate") to Gal-Cer, by the microsomal sulfotransferase.

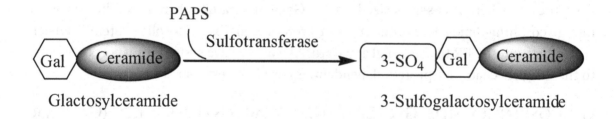

Ganglioside Synthesis

Most gangliosides are synthesized from lactosylceramide in the lumen of the Golgi apparatus by glycosyltransferases and the stepwise addition of activated monosaccharides and sialic acid, usually N-acetylneuraminic acid (NANA or NeuAc). However, the minor membrane ganglioside known as GM4 is produced from galactosylceramide (Gal-Cer). The activated sugars involved are; UDP-glucose (UDP-Glc), UDP-galactose (UDP-Gal), and UDP-N-acetylgalactosamine (UDP-GalNAc).The first product is the ganglioside GM3, which is formed by sialyltransferase I.

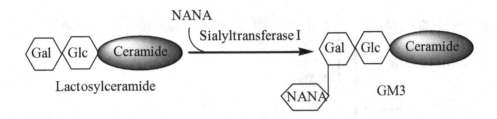

GM3 is converted to GM2 by a specific glycosyltransferase.

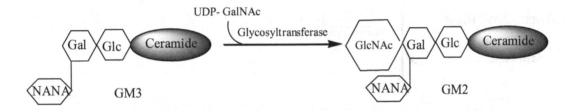

GM1 is formed by a specific glycosyltransferase.

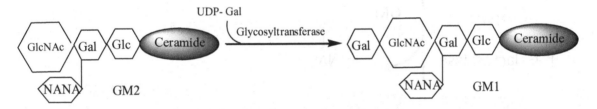

DEGRADATION OF GLYCOSPHINGOLIPIDS

Degradation of glycosphingolipids occurs in lysosomes and requires a group of hydrolytic enzymes. Figure 6.1 shows ganglioside GM1degradation. In this process, specific glycosidases sequentially cleave off the terminal carbohydrate residues. For glycolipids with short oligosaccharide chains, membrane-active sphingolipid activator proteins (SAPs) are also required. The glycosphingolipids must first be internalized into cells by endocytosis so that lysosomes can fuse with the endocytotic vesicles. The lysosomal enzymes includes; α- and ß-galactosidases, ß-glucosidase, neuraminidase, hexosaminidase, sphingomyelinase, ceramidase, and sulphatase. The last residue added during synthesis is the first to be removed. Thus, ganglioside GM1 is degraded by ß-galactosidase into ganglioside GM2 and galactose. Then GM2 is cleaved by β-hexosaminidase A to GM3 and N-acetyl-galactosamine (GlcNAc).

Figure 6.1

Ganglioside GM1degradation. Ganglioside GM1 is first degraded by ß-galactosidase into ganglioside GM2 and galactose. Then GM2 is cleaved by β-hexosaminidase A to N-acetyl-galactosamine (GlcNAc) and GM3, which is further degraded into sphingosine and free fatty acid.

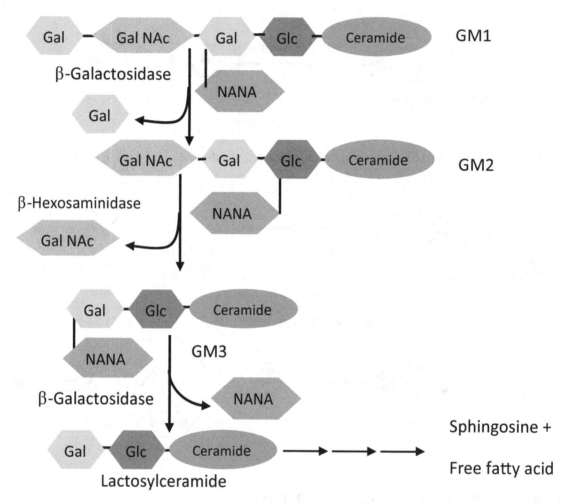

LIPID-STORAGE DISEASES

These storage diseases, also called sphingolipidosis, are inherited disorders in which, typically, specific lipids accumulate in cells and tissues. They are caused by defects in genes that encode inactive or poorly active enzymes for the degradation of sphingolipids. The lipid that accumulates within the lysosome is the substrate for the deficient enzyme. Selected examples of the more common sphingolipidoses are summarized in table 6.1.

All are inherited in an autosomal-recessive fashion with the exception of Fabry's disease, which is X-linked. They are frequent among patients of Ashkenazi Jewish origin. Diagnosis is made by examination of tissue samples, cultured fibroblasts or peripheral leukocytes, and amniotic fluid for the presence of the degradative enzymes and for the accumulated lipids.

Table 6.1

Some common sphingolipidoses.

Disease	Enzyme Deficiency	Accumulated Lipid
GM$_1$	β-Galactosidase	GM$_1$ ganglioside
Tay-Sachs disease	β-Hexosaminidase A	GM$_2$ ganglioside
Fabry's disease	α-Galactosidase A	Trihexosylceramide
Gaucher's disease	β-Glucosidase	Glucosylceramide
Niemann-Pick disease type A	Sphingomyelinase	Sphingomyelin
Farber's disease	Ceramidase	Ceramide
Krabbe's disease	β-Galactosidase	Galactosylceramide
Metachromatic leuko-dystrophy	Arylsulfatase A	Sulfatides
Sandhoff's disease	Hexosaminidase A and B	GM$_2$ and globoside

TREATMENT OF SPHINGOLIPIDOSES

Despite determined efforts to find treatment for the sphingolipidoses, only a few strategies are available. These include gene therapy, enzyme replacement therapy, enzyme stabilization, bone-marrow transplantation, substrat- reduction therapy, and hematopoietic stem-cell transplantation. Bone-marrow transplantation can cure patients with type 1 Gaucher disease. However, because of the risk associated with this procedure, only a limited number of patients have received this treatment. Patients with type 3 Gaucher disease have received this treatment with variable clinical benefits. Enzyme replacement therapy is highly effective for patients with type 1 Gaucher disease. Patients showed improvement of their anemia, thrombocytopenia, hepatosplenomegaly, and skeletal damage and in the quality of their lives. Type 3 Gaucher disease patients given the same treatment showed hematologic and systemic manifestations.

CONCLUSION

- Glycosphingolipids, or glycolipids, consist of a carbohydrate moiety linked to ceramide. They include cerebrosides, globosides, sulphatides, and gangliosides, all of which are essential components of cell membranes and important constituents of nervous tissue.

- Their synthesis starts with ceramide and takes place in the endoplasmic reticulum and Golgi apparatus. Cerebrosides contain one unit of glucose or of galactose. Globosides contain two or more carbohydrate units. Sulphatide is galacosyl cerebroside-3-sulphate. Gangliosides are lactosylceramides to which other monosaccharide units and one or more N-acetylneuraminic acid units are linked.
- Their stepwise degradation in lysosomes requires specific hydrolytic enzymes. Congenital deficiency of one of these results in a specific sphingolipidosis (lipid-storage disease).

SELECTED READINGS

Gault, C. R., Obeid, L. M., and Hannun, Y. A. (2010). An overview of sphingolipid metabolism: From synthesis to breakdown. *Advances in Experimental Medicune and Biology* 688: 1–23.

IUPAC-IUB Joint Commission on Biochemical Nomenclature (JCBN). (1998). Nomenclature of glycolipids. Recommendations. *European Journal of Biochemistry* 257: 293–298.

Moser, H. W., Smith, K. D., and Moser, A. B. (1995) X-linked adrenoleukodystrophy. In C. R. Scriver, A. L. Beaudet, W. S. Sly, et al. (Eds.), *The metabolic and molecular basis of inherited disease*, 7th Ed.,1Vol. II. New York: McGraw-Hill, 2325–2424.

Sandhoff, K., and Kolter, T. (2003). Biosynthesis and degradation of mammalian glycosphingolipids. *Philosophical Transactions of the Royal Society* 358 (1433): 847–861.

Schulze, H., and Sandhoff, K. (2011). Lysosomal lipid storage diseases. *Cold Spring Harbor Perspectives in Biology* doi: 10.1101/cshperspect.a004804.

Watts, W. E., and Gibbs, D. A. (1986). Lysosomal storage diseases: Biochemical and clinical aspects. London: Taylor and Francis, 158–161.

Vance, D. E., and Vance, J. E. (Eds.). (1985). *Biochemistry of lipids and membranes*. Menlo Park, CA: Benjamin/Cummings, 25–72.

Chapter 7: Metabolism Of Cholesterol

Introduction

For many years there has been strong epidemiologic evidence of continuous association between the risk of coronary heart disease (CHD) and elevated plasma cholesterol levels. However, most people are unaware that cholesterol plays crucial roles in the human body, including the control of cell-membrane fluidity and as a precursor of steroid hormones, bile salts, and vitamin D_3. The cholesterol present in the human body is derived from the diet and from de novo synthesis. The latter decreases when the diet provides a sufficient amount of cholesterol; however, low cholesterol in the diet stimulates de novo synthesis from acetyl-CoA in all body tissues, with the liver accounting for approximately half of the total synthesis. Rich dietary sources of cholesterol include dairy products, egg yolks, and meat, including liver.

Cholesterol is transported in blood by plasma lipoproteins. From the small intestine, dietary cholesterol enters the blood in the form of chylomicrons (CM). Very-low-density lipoprotein (VLDL) carries newly synthesized cholesterol from hepatocytes to the blood and is there converted into low-density-lipoprotein (LDL), which delivers cholesterol to peripheral tissues. High-density lipoprotein (HDL) transports cholesterol from extra-hepatic tissues to the liver. Cholesterol is eliminated from the human body unchanged or in the form of bile salts. Secretion of cholesterol into bile in excessive quantities leads to gallstone formation, and the deposition of cholesterol in arteries leads to atherosclerosis.

This chapter discusses the biosynthesis of cholesterol, its degradation, and its excretion. The transport of cholesterol and its role in the development of atherosclerosis and coronary heart disease will be discussed in chapter 8.

Biosynthesis of Cholesterol

All the carbon atoms in cholesterol are synthesized from acetyl-CoA by all body tissues. A lot of recent evidence has indicated that peroxisomes contain enzymes for cholesterol synthesis that were previously thought to be located in cell cytosol or endoplasmic reticulum, such as acetoacetyl-CoA thiolase, HMG-CoA synthase, HMG-CoA reductase, mevalonate kinase, phosphomevalonate kinase, phosphomevalonate decarboxylase, isopentenyl diphosphate isomerase, and farnesyl diphosphate synthase. Cholesterol synthesis may be divided into three stages: (1) formation of mevalonate, (2) formation of squalene, and (3) conversion of squalene to cholesterol.

1. Formation of mevalonate

The initial reaction in this process is the formation of acetoacetyl-CoA from two molecules of acetyl-CoA by acetoacetyl-CoA thiolase. Condensation of a third molecule of acetyl-CoA by HMG-CoA synthase produces 3-hydroxy-3-methylglutaryl- CoA (HMG-CoA). The HMG-CoA is then reduced to mevalonate by HMG-CoA reductase, the rate-limiting enzyme and the most highly regulated step in cholesterol biosynthesis. Figure 7.1 shows these reactions.

Figure 7.1

Biosynthesis of mevalonate.

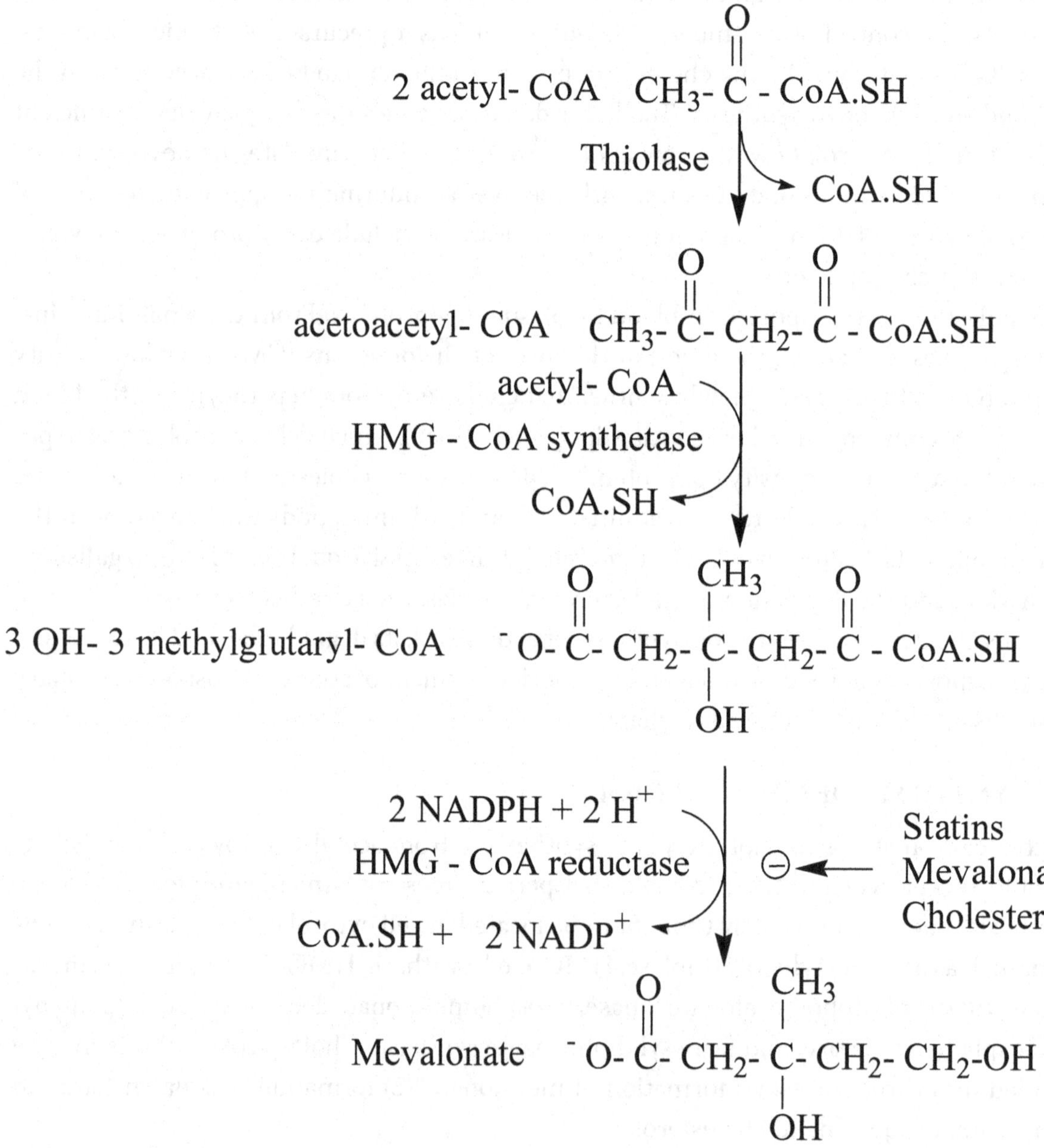

2. Formation of squalene

The second stage is responsible for the formation of squalene from mevalonate. This starts with the production of isopentenyl pyrophosphate and dimethylallyl pyrophosphate (see figure 7.2). This involves activation of mevalonate by three successive phosphorylations to produce 3-phospho-5-pyrophosphomevalonate, which is converted by pyrophosphomevalonate decarboxylase to isopentenyl pyrophosphate.

Then, isopentenyl pyrophosphate is isomerized to dimethylallyl pyrophosphate by isopentenyl pyrophosphate isomerase. Condensation of dimethylallyl pyrophosphate with isopentenyl pyrophosphate by dimethylallyl transferase yields geranyl pyrophosphate. The yield reacts with isopentenyl pyrophosphate to produce farnesyl pyrophosphate by action of geranyl transferase. Finally, squalene results from condensation of two molecules of farnesyl pyrophosphate by farnesyl transferase, which is also called squalene synthase. Figure 7.3 shows the formation of squalene from isopentenyl pyrophosphate and dimethyl-allyl pyrophosphate.

3. Conversion of squalene to cholesterol

The last stage in cholesterol biosynthesis (see figure 7.4) starts with the binding of squalene to sterol carrier protein (a cytosolic protein). Squalene is converted to lanosterol in two steps. Squalene-monooxygenase introduces an epoxide function at carbons 2 and 3 to yield squalene 2,3-epoxide, which is then converted to lanosterol by 2,3-epoxidosqualene cyclase. Both enzymes involved in the formation of lanosterol have been localized in the endoplasmic reticulum. Cholesterol is formed from lanosterol through a series of about 20 reactions in the endoplasmic reticulum membrane. These involve the reduction of double bonds and 3 demethylations to produce 7-dehydrocholesterol, which is reduced by NADPH to cholesterol.

REGULATION OF HMG-CoA REDUCTASE

The activity of this enzyme is regulated by three mechanisms:

1. At the level of gene expression, the production of mRNA is modulated by the supply of cholesterol. A low intracellular cholesterol level stimulates synthesis of the mRNA for the enzyme and increases cholesterol synthesis. In contrast, a high intracellular cholesterol level inhibits synthesis of this mRNA and decreases cholesterol synthesis. This may be the main regulatory mechanism and involves the sensing of intracellular cholesterol in the endoplasmic reticulum by the sterol-regulatory-element binding proteins (SREBP 1 and 2). SREBP 2 activates transcription of the gene for the low-density lipoprotein receptor and of genes for cholesterol biosynthesis; whearse SREBP 1 is required in the regulation of the *de novo* synthesis of fatty acids. In the presence of cholesterol, SREBP2 binds SCAP (SREBP-cleavage activating protein) and Insig-1. Insig-1 and Insig-2 are ER-membrane proteins encoded by insulin-inducible genes 1 and 2. In cholesterol-deprived cells, Insig-1 dissociates from the SREBP-SCAP complex, allowing the complex to migrate to the Golgi apparatus. S1P and S2P (site-1 and -2 protease), two enzymes that are activated by SCAP when cholesterol levels are

low, they release the NH_2 -terminal of SREBP. The cleaved SREBP then enters the nucleus and activates target-gene expression. When sterol accumulates, the membrane domain of the SCAP binds to one of the two ER membrane proteins, Insig-1and Insig-2, in the ER membrane. The binding of this protein blocks the migration of SREBP to the nucleus and prevents the proteolytic activation of SREBP. Expression of the SREBP target genes then declines and consequently suppresses cholesterol synthesis and uptake. Michael Brown and Joseph Goldstein, Nobel Prize winners for medicine in 1985 for their excellent work on the regulation of cholesterol metabolism, made great contributions to this pathway.

2. HMG-CoA reductase is regulated through its degradation, which increases when cholesterol accumulates in the cell and decreases when there is less cholesterol.

3. HMG-CoA reductase is also regulated through phosphorylation and dephosphorylation cycle, which deactivate and activate the enzyme, respectively (see figure 7.5). Cholesterol synthesis is enhanced by insulin, which decreases the formation of cAMP and prevents activation of HMG-CoA reductase kinase.. The HMG-CoA reductase remains unphosphorylated and therefore active. In contrast, glucagon enhances formation of cAMP and inhibits the activity of HMG-CoA reductase.

Statins (such as atorvastatin, lovastatin, mevastatin, simvastatin, and pravastatin) are drugs used to treat hypercholesterolemia. They selectively inhibit HMG-CoA reductase.

DEGRADATION OF CHOLESTEROL

Bile acid formation is the main route of cholesterol catabolism. Bile salts are amphipathic molecules that are powerful physiological detergents for absorption and transport of lipids and fat-soluble vitamins. Cholesterol is converted to bile acids in the liver, which secretes the acids into the bile for transport into the intestine, where they function or are eliminated. In the lower intestine, bacterial flora convert cholesterol into neutral fecal sterols, especially coprostanol. Approximately l g of cholesterol is eliminated from the body, and about half of it is excreted in the feces as bile acids per day.

SYNTHESIS OF BILE ACIDS

The primary bile acids (cholic acid and chenodeoxycholic acid) are synthesized in the liver by a multistep pathway in which hydroxyl groups are inserted at specific positions on the steroid structure, the double bond of the B ring is reduced, and 3 carbons are removed from the end of the hydrocarbon chain and replaced by a carboxyl group (see figure 7.6). Then, 7α-hydroxylase converts cholesterol to 7α-hydroxycholesterol. Its activity and its expression are inhibited by bile acids. In the liver, the bile acids are conjugated with glycine or taurine through an amide bond between the carboxyl group of the bile acid and the amino group of glycine or taurine to produce bile salts. These include glycocholic acid, glycochenodeoxycholic acid, taurocholic acid, and taurochenodeoxycholic acid. The secondary bile acids (deoxycholic acid and lithocholic acid) are formed by intestinal bacteria from primary bile acids in a process involving deconjugation and 7α-dehydroxylation (see figure 7.6).

Figure 7.2

Formation of isopentenylpyrophosphate and dimethylallylpyrophosphate.

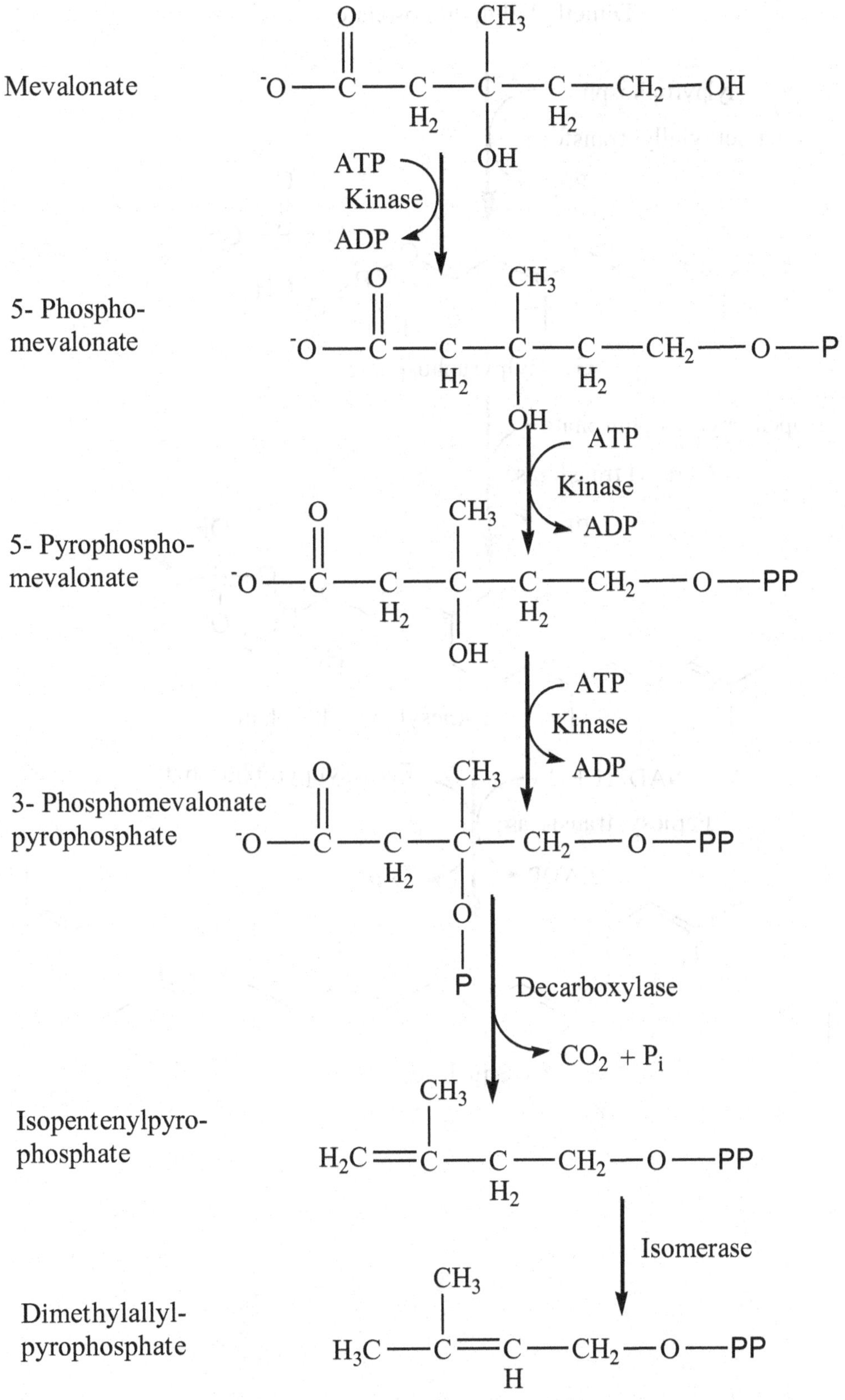

Figure 7.3

Squalene biosynthesis from isopentenyl pyrophosphate and dimethylallyl pyrophosphate.

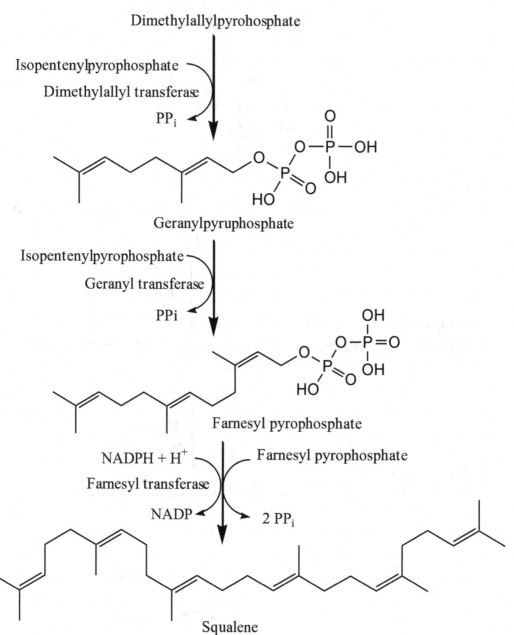

Figure 7.4

Biosynthesis of cholesterol from squalene.

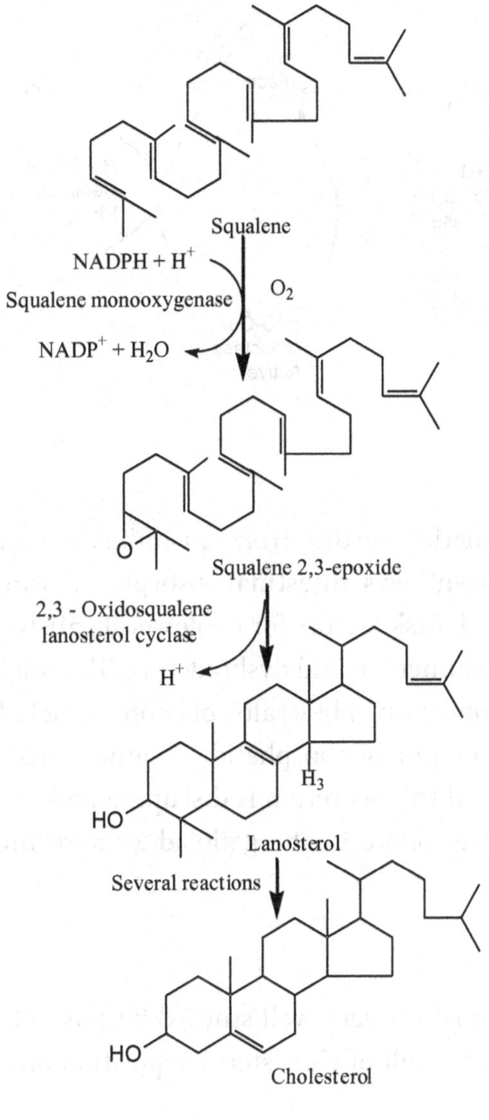

ENTEROHEPATIC CIRCULATION

Most of the bile acids secreted into the upper small intestine are absorbed in the lower small intestine and returned to the liver via portal circulation, also known as the enterohepatic circulation, which is a tightly regulated cycle. Less than 5% of secreted bile acids escape intestinal reabsorption and excretion via feces, but this amount needs to be replaced by de novo synthesis to keep the level of bile acids constant. Enterohepatic bile-acid circulation is vital for several liver and gastrointestinal functions, including bile flow, solubilization and excretion of cholesterol, clearance of toxic molecules, intestinal absorption of lipophilic nutrients, and metabolic and antimicrobial effects.

The bile acids are transported in the portal blood in complexes with albumin.

Figure 7.5

Control of HMG-CoA reductase activity by phosphorylation and dephosphorylation.

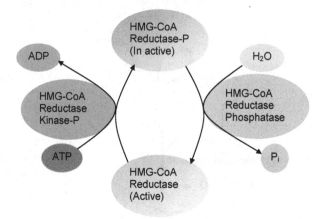

CHOLELITHIASIS

Cholesterol-gallstone formation results from an imbalance in cholesterol metabolism, which involves hepatic biosynthesis, intestinal absorption, formation of bile salts, and biliary excretion of cholesterol. Risk factors for cholesterol gallstones include obesity, type 2 diabetes mellitus, hyperinsulinemia, and dyslipidemia. The solubility of cholesterol in bile depends on the relative proportion of bile salts, phosphatidylcholine, and cholesterol itself. Therefore, simultaneous secretion of phosphatidylcholine and bile salts into bile is required to keep cholesterol soluble. If this secretion is disrupted and more cholesterol is delivered to bile, cholesterol may precipitate in the gallbladder and initiate cholesterol-gallstone formation (cholelithiasis).

CONCLUSION

- Cholesterol metabolism is very well studied because of its association with coronary heart disease, a result of cholesterol deposition on the inner walls of arteries, or atherosclerosis.
- The de novo synthesis of the 27-carbon cholesterol from acetyl-CoA takes place through a complex series of reactions in which 5-carbon isoprene units are conjugated to form the 30-carbon squalene, followed by cyclization and modification to the desired product.
- Cholesterol is transported in blood by plasma lipoproteins, chylomicrons, very-low-density lipoprotein, low-density lipoprotein, and high-density lipoprotein.
- It is eliminated from the human body as cholesterol or in the form of bile salts. Aggregation of cholesterol in bile leads to gallstone formation.
- Cholesterol is the precursor of hormones of the adrenal cortex, testes, and ovaries; of bile acids in the liver; and of vitamin D in the skin.

Figure 7.6

Biosynthesis of bile acids.

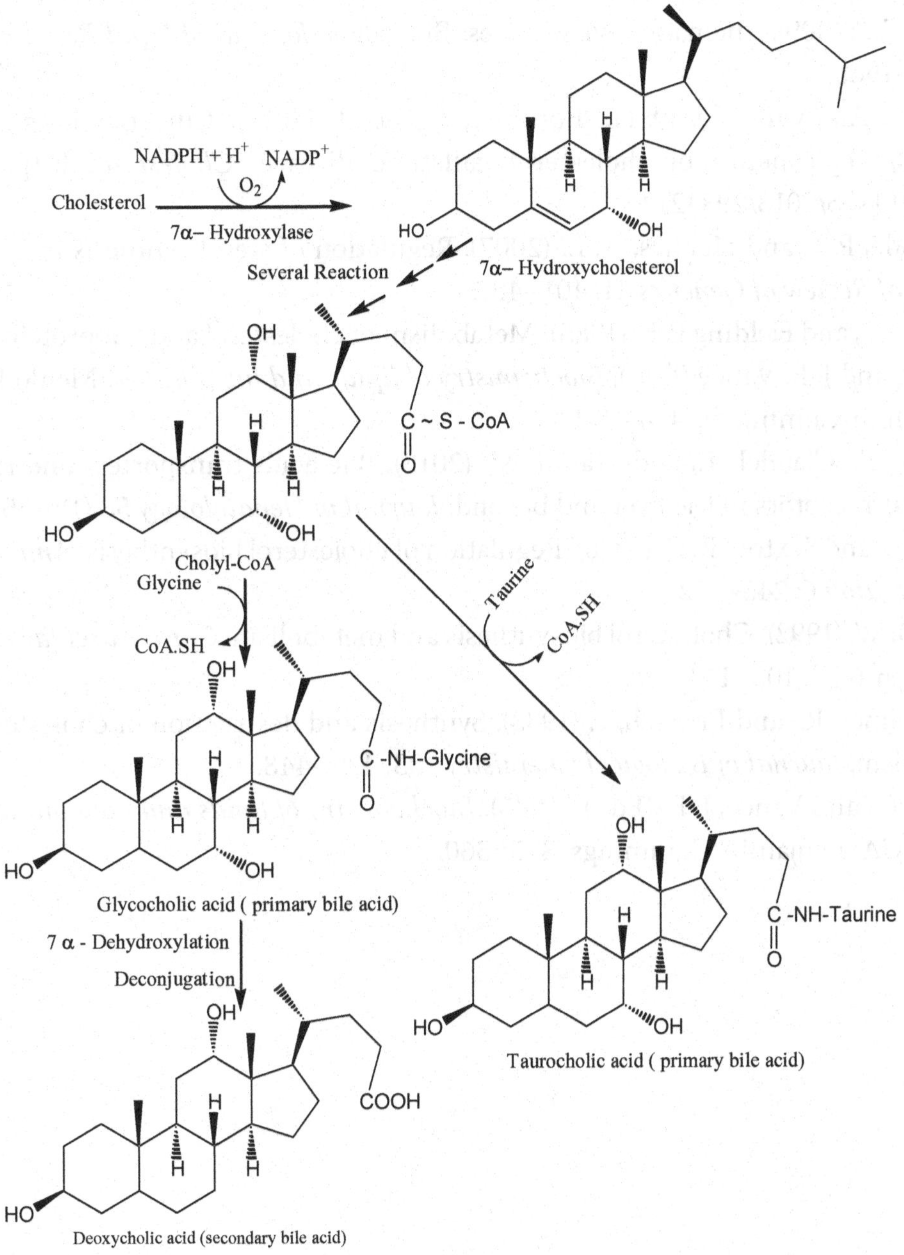

SELECTED READINGS

Brown M. S., and Goldstein, J. L. (1980). Multivalent feedback regulation of HMG- CoA reductase, a control mechanism coordinating isoprenoid synthesis and cell growth. *The Journal of Lipid Research* 21: 505–517.

Brown M. S., and Goldstein, J. L. (1997). The SREBP pathway: Regulation of cholesterol metabolism by proteolysis of a membrane-bound transcription factor. Cell 89; (3): 331–34.

Brown M. S., and Goldstein, J. L. (1999). A proteolytic pathway that controls the cholesterol content of membranes, cells, and blood. *Proceedings of the National Academy of Sciences of the USA* 96: 11041–11048.

Chiang J. L. (2009). Thematic review series: Bile acids. *Journal of Lipid Research* 50 (10): 1955–1966.

Di Ciaula, A., Wang, D. Q.-H., Bonfrate, L., et al. (2013). Current views on genetics and epigenetics of cholesterol gallstone disease. *Cholesterol* http: //dx.doi.org/10.1155/2013/298421.

Espenshade, P. J., and Hughes, A. L. (2007). Regulation of sterol synthesis in eukaryotes. *Annual Review of Genetics* 41: 401–407.

Fielding C. J., and Fielding P. E. (1985). Metabolism of cholesterol and lipoproteins. In D. E. Vance and J. E. Vance (Eds.), *Biochemistry of lipids and membranes.* Menlo Park, CA: Benjamin/Cummings, 404–474.

Halilbasic, E., Claudel, T., and Traunr, M. (2013). Bile acids transporters and regulatory nuclear receptors in the liver and beyond. *Journal of Hepatolology* 58 (1): 155–168.

Rudney, H., and Sexton, R. C. (1986). Regulation of cholesterol biosynthesis. *Annual Review of Nutrition* 6: 245–272.

Russell, D. W. (1992). Cholesterol biosynthesis and metabolism. *Cardiovascular Drugs and Therapy* 6 (2): 103–110.

Schoenheimer, R., and Breusch, F. (1933). Synthesis and destruction of cholesterol in the organism. *Journal of Biological Chemistry* 103, 439–448.

Vance D. E., and Vance, J. E. (Eds.). (1985). *Biochemistry of lipids and membranes.* Menlo Park, CA: Benjamin/Cummings, 325–360.

Chapter 8: Metabolism of Lipoproteins

Introduction

Since lipids are insoluble in an aqueous medium such as plasma, they require a carrier to be transported through such a medium. This problem is resolved by the presence of plasma lipoproteins, which are complexes of a variety of lipids with one or more specific proteins called apolipoproteins (apoproteins). The lipid and protein are held together by noncovalent forces in spherical particles that contain a central core of a neutral lipid (of triacylglycerol [TAG], cholesterol ester, or both) surrounded by a layer of polar lipids, primarily phospholipids, with the polar head groups pointing outwards and one or more apoproteins inserted into this outer shell (figure 8.1). Free cholesterol is present mainly in the surface monolayer, but with increasing particle size, it is progressively distributed into the central core (figure 8.1). There is an inverse correlation between plasma concentration of high-density lipoprotein cholesterol (HDL-C) and its major protein, apolipoprotein A-I (apo A-I) with atherosclerotic cardiovascular disease risk in humans. In contrast, high concentration of plasma low-density lipoprotein cholesterol (LDL-C) has been positively correlated with increased incidence of atherosclerosis and cardiovascular diseases.

Types of Lipoprotein

Human plasma lipoproteins have been classified into 5 major classes on the basis of their density, which is inversely related to their size, as determined by ultracentrifugation and electrophoresis. Table 8.1 shows normal levels in plasma, while table 8.2 shows the apoprotein content of the various lipoproteins.

Table 8.1

Normal levels of human-serum lipids and lipoproteins (Third Report of the National Cholesterol Education Program, NCEP, Expert Panel, ATPIII 2002).

Lipid	Concentration
Total cholesterol	< 200 mg % Desirable
HDL – cholesterol	> 40 mg %
LDL – cholesterol	< 100 mg % Optimal
Triglycerides	< 150 mg %

Chylomicrons (CMs) are the lowest in density (< 0.95 g/ml) and largest in size (with a diameter of 75 to 1200 nm) and contain the most lipid (98 to 99% of dry weight, mainly TAG) and the least protein (1 to 2% of dry weight).

Table 8.2

The apoprotein composition of plasma lipoproteins (HDL: high-density lipoprotein, IDL: intermediate-density lipoprotein, LDL: low-density lipoprotein, VLDL: very-low-density lipoprotein).

Lipoprotein	Apolipoprotein
CMs	A-I, A-II, A-IV, B-48, CI, C-II, C-III
VLDL	B100, C-I, C-II, C-III, E
IDL	B100, C-I, C-II, C-III, E
LDL	B- 100, C-III, E
HDL	A-I, A-II, C- I, C-II, C-III, D, E, J, L-I, M

HDL (d = 1.0063–1.210 g/ml) is the smallest and contains the most protein (33% of dry weight) and the least lipid (67% of dry weight, rich in cholesterol ester).

Figure 8.1

General structure of plasma lipoprotein. Lipoprotein particles are composed of a lipid core containing cholesteryl esters and triglycerides, and a surface coat of phospholipids, unesterified cholesterol and apolipoproteins. Used with permission from AntiSense, (2010), Wikipedia Commons, the free media repository.

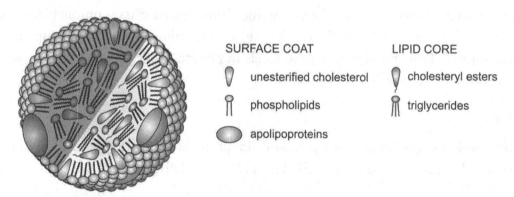

LDL (d = 1.019–1.063 g/ml), IDL (d =1.006–1.019 g/ml), and VLDL (d < 1.006 g/ml) are between CMs and HDL in size and composition. LDL and VLDL are rich in cholesterol and TAG, respectively. About two-thirds of the total plasma cholesterol (free plus esters) is contained in LDL. Each type of lipoprotein has a characteristic apolipoprotein composition.

Apolipoproteins are proteins that bind hydrophobic lipids in the blood and help solubilize them. There are two major types of apoproteins: the non-exchangeable type, and the exchangeable type. The non-exchangeable type consists of apolipoprotein B, which occurs in plasma in 2 isoforms (apo B-100 and apo B-48), each of which is tightly integrated into the phospholipid monolayer of VLDL and CM, respectively. Both isoforms are encoded by the same gene and by a single mRNA transcript larger than 16 kb. Apo B-48 is produced by the small intestine when a stop codon (UAA) at residue 2153 is created by editing of the mRNA. In fact, apo B-48 is so called because it constitutes only 48% of the apo B-100 sequence. Apo B-100 is a ligand for the LDL receptor required for VLDL production by the liver. The exchangeable type of apolipoproteins are less tightly associated with the phospholipid monolayer and exchange among the lipoproteins such as apo C and apo E. The functions of these proteins are given in table 8.3. Lipoproteins are in a constant state of synthesis, degradation, and removal from the circulation.

METABOLISM OF CHYLOMICRONS

CMs are formed in intestinal mucosal cells and transport dietary lipid (TAG, cholesterol, and cholesterol esters) from the intestine to other tissues in the human body (figure 8.2).

Table 8.3

Major function of the human plasma apolipoproteins.

A- I	The major protein in HDL; activates lecithin-cholesterol acyltransferase (LCAT)
A- II	Regulates HDL stability and affects hepatic lipase association and activity
B-100	The major protein in LDL; ligand for LDL receptor
B- 48	Found mainly in chylomicrons
C-I	Inhibitor of cholesterol ester transfer protein & inhibitor of lipoprotein binding to LDL receptor
C-II	Activator of extra hepatic lipoprotein lipase
C-III	Inhibitor of lipoprotein lipase
D	Found on HDL; significant player in atherosclerosis.
E	Ligand for IDL, chylomicron remnant receptors in liver

FORMATION OF CHYLOMICRONS

CMs are produced and assembled in the endoplasmic reticulum (ER) of the enterocytes that line the intestinal lumen. Their assembly occurs during transfer of their components from the ER to the Golgi apparatus, where they are packaged into secretory vesicles and delivered by exocytosis into the extracellular space as nascent CMs rich in TAG formed from absorbed dietary fatty acids. Their protein is synthesized in enterocytes and includes apo B-48 and apo A-I, A-II, and A-IV. After transfer from the thoracic duct into the blood stream, the nascent CMs are rapidly converted into mature CMs by transfer of apo E, (the ligand for LDL and CM receptors on hepatocytes) and apo C-II from HDL in the plasma.

CMs are initially degraded by action of lipoprotein lipase, an extracellular enzyme found on the capillary endothelial walls of extrahepatic tissues. This lipase, in the presence of apo C-II, hydrolyzes most of the TAG into monoacylglycerol (MAG), free fatty acids, and glycerol. The decrease in TAG of the CMs results in the return of apo C-II to HDL. The CM remnants (rich in cholesterol) are taken up by hepatocytes in a process mediated by a receptor specific for apo E, and their apolipoproteins and cholesterol esters are degraded by lysosomal enzymes into amino acids, cholesterol, and fatty acids. The cholesterol in the liver can be converted into bile acids and secreted in bile as such, incorporated into VLDL, or converted into cholesterol esters. Cholesterol also regulates the rate of de novo synthesis of cholesterol in the liver by inhibiting activity and synthesis of HMG-CoA reductase.

METABOLISM OF VERY-LOW-DENSITY LIPOPROTEIN (VLDL)

The formation of VLDL in hepatocytes is similar to that for CMs in enterocytes. It transports excess TAG (derived from CM remnants, lipogenesis, or plasma free fatty acids) from the liver to extrahepatic tissues, and it also transports cholesterol esters to the peripheral tissues. The released nascent VLDL (containing apo B-100 and A-I) receives apo C-II and apo E from HDL in the plasma. Most TAG of VLDL is degraded by lipoprotein lipase, as described for chylomicrons. This causes the VLDL to decrease in size, become dense, form a remnant VLDL called IDL, and permit apo C-II to be transferred back to HDL. The IDL is taken up by hepatocytes in a process mediated by the LDL (apo B-100, apo E) receptor or is converted into LDL after further hydrolysis by lipoprotein lipase and enrichment in cholesterol ester by transfer from HDL by cholesterol-ester transfer protein. LDL is taken up by hepatocytes via apo E receptors or by extrahepatic tissues via apo B-100 receptors. Figure 8.3 shows the metabolism of VLDL.

METABOLISM OF LOW-DENSITY LIPOPROTEIN (LDL)

LDL is formed from VLDL in the plasma. It is the lipoprotein richest in cholesterol, constituting about two-thirds of the total cholesterol in the plasma. LDL transports cholesterol to peripheral tissues by depositing free cholesterol on cell membranes when LDL comes into contact with it, and by binding to specific receptors (figure 8.4) that recognize apo B-100. Low-density lipoprotein receptor (LDL-R) is a transmembrane glycoprotein that mediates endocytosis of LDL by recognizing apo B-100 in the outer layer of LDL. LDL-R also recognizes apoprotein E on chylomicron and VLDL (IDL) remnants. LDL-R is polypeptide of 839 residues organized into five structural domains. Figure 8.4 shows the sequential steps in the LDL-receptor pathway of mammalian cells.

Apo B-100 binds LDL-R that congregates in areas of the plasma membrane called coated regions or pits, which are coated with the protein clathrin. Pits loaded with LDL form coated vesicles, which detach from the membrane and enter the cell by endocytosis.

The entry of protons through the membrane of these endosomes increases their acidity and causes the LDL to dissociate from its receptors. The LDL-R that remains in the endosomes can be recycled back to the cell membrane, whereas the free lipoprotein is degraded by lysosomal enzymes to amino acids, fatty acids, phospholipids, and free cholesterol. The released cholesterol exerts three regulatory effects: First, it decreases de novo synthesis of cholesterol by inhibiting HMG-CoA reductase activity through suppression of transcription of this enzyme's gene. Second, it promotes the storage of excess cholesterol in the cell by enhancing the formation of cholesterol esters from cholesterol and a long-chain acyl-CoA through activation of acylcholesterol acyltransferase (ACAT). Third, it reduces the synthesis of LDL receptors by suppressing the transcription of the LDL gene, thereby limiting the entry of cholesterol into the cell.

$$\text{Cholesterol + Acyl-CoA} \xrightarrow{\text{ACAT}} \text{Cholesterol-ester + CoA.SH}$$

Most of the cholesterol released from LDL moves to the endoplasmic reticulum to be used for membrane synthesis.

Figure 8.2

Metabolism of chylomicrons. Nascent CMs are converted in plasma into mature CM by transfer of Apo C-II and Apo E from HDL. Hydrolysis of most of their TAG by lipoprotein lipase yields CM remnants which are taken up via a hepatocyte receptor that recognizes Apo E. (Apo A, apolipoprotein A; Apo B-48, apolipoprotein B-48; Apo B-100, apolipoprotein B-100; Apo C-II, apolipoprotein CII; Apo E, apolipoprotein E; CM, chylomicrons; CM-R; chylomicron remnants N-CM, nascent chylomicrons).

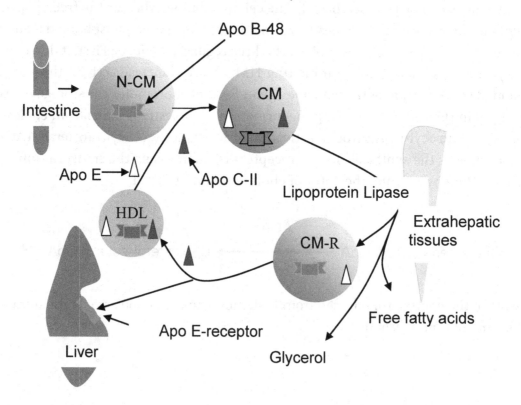

Figure 8.3

Metabolism of VLDL. Nascent VLDL is converted into VLDL after acquiring apo C-II and apo E from HDL. IDL is formed when most of the VLDL TAG is hydrolyzed by lipoprotein lipase. IDL is converted into LDL when more TAG hydrolysis occurs and apo C-II is returned to HDL. (Apo A: apolipoprotein A, apo B-48: apolipoprotein B-48, apo B-100: apolipoprotein B-100, apo C-II: apolipoprotein C-II, apo E: apolipoprotein E, HDL: high-density lipoprotein, VLDL: very-low-density lipoprotein, N-VLDL: nascent-VLDL, LDL: low-density lipoprotein).

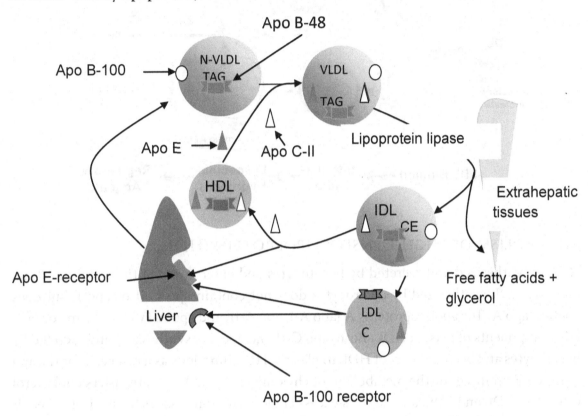

Figure 8.4

Cellular uptake of low-density lipoprotein (LDL). Used with permission from Brown, M. S., and Goldstein, J. L. (1979). Receptor mediated endocytosis: Insights from the lipoprotein receptor system. *Proceedings of the National Academy of Science, USA 76:3330–3337.*

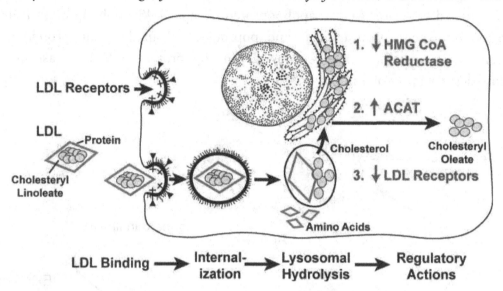

METABOLISM OF HIGH-DENSITY LIPOPROTEIN (HDL)

HDL is synthesized and secreted by hepatocytes and enterocytes of the small-intestinal mucosa. HDL synthesized by enterocytes does not contain apo C-II or apo E but does contain apo A. The apolipoproteins A (apo A-I, apo A-II, and apo A-IV) are the major protein components of plasma HDL. Both, apo C-II and apo E is synthesized and secreted by hepatocytes and becomes part of HDL in plasma. HDL functions as a reservoir for the apo C-II and E required for the metabolism of chylomicrons and VLDL, transfers cholesterol esters to VLDL and LDL, and delivers excess cholesterol from tissues to the liver, where it is excreted as bile salts into the intestine.

The protective effect of HDL against atherosclerosis is usually attributed to its role in reverse cholesterol transport, whereby HDL promotes the flux of excess cholesterol from peripheral tissues to the liver for excretion. HDL maturation begins with secretion of lipid-poor apo A-I by the liver and intestine. Cholesterol and phospholipids released by the liver and cholesterol, phospholipids, and apolipoproteins from chylomicrons and VLDL are then acquired to form nascent pre-β-HDL particles. Cholesterol and phospholipids are also acquired from extrahepatic tissues so that the particles become progressively enriched with cholesterol. The LCAT on HDL particles esterifies the free cholesterol to cholesteryl ester, which migrates to the core of the HDL particles.

Figure 8.5 shows the secretion, lipid acquisition, and maturation of HDL particles. Newly secreted HDLs (disc shaped) are good acceptors of cholesterol from the surface of cell membranes and from other plasma lipoproteins. They become spherical particles by the accumulation of cholesterol esters formed from cholesterol and phosphatidylcholine (lecithin) on the surface of HDL by lecithin-cholesterol-acyl-transferase (LCAT). In plasma, this enzyme is found bound to a species of HDL particles and is activated by apo A-I and inhibited by apo C-III. The lysolecithin product of the reaction is bound in plasma by albumin, whereas the cholesterol esters are transferred from HDL to VLDL or LDL by the cholesterol ester transfer protein (CETP). The spherical particles are taken up by hepatocytes via the scavenger receptor. This receptor mediates the last step in cholesterol reverse transport, or selective cholesterol ester uptake: the delivery of cholesterol esters from the core of HDL to the liver without degradation of HDL via a receptor that binds to the scavenger receptor with apo A-I.

The smaller HDL particles, HDL3, then accept unesterified cholesterol from tissues via ATP-binding cassette transporter 1 (ABC-1). The cholesterol content of HDL3 is converted into cholesterol ester by LCAT, leading to the generation of the larger and less dense HDL particle known as HDL2, which is taken up by the liver via the scavenger receptor. In the liver, the released cholesterol ester is converted to free fatty acid and free cholesterol. The free cholesterol can be excreted into the bile after its conversion to bile salts.

Figure 8.5

Secretion, lipid acquisition, and maturation of HDL particles. Used with permission from Lippincott Williams and Wilkins/Wolters-Kluwer Health (Lewis and Rader 2005).

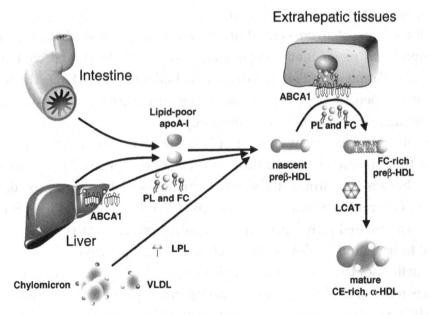

FATTY LIVER

Nonalcoholic fatty liver disease (NAFLD)

NAFLD comprises a spectrum of disorders that range from simple steatosis (fatty liver) through nonalcoholic steatohepatitis (NASH) to fibrosis and ultimately to liver cirrhosis. Lipid (mainly TAG) accumulation in the liver due to abnormalities in lipid and lipoprotein metabolism is the hallmark of NAFLD. Fatty liver is associated with increased plasma LDL and triglycerides and decreased HDL. The association of fatty liver and the concentration of small, dense LDL (sdLDL) is well established.

As the triglyceride-rich VLDL enters plasma at an accelerated rate, sdLDL, the most atherogenic subclass of LDL, develops as triglycerides are gradually removed from VLDL. Production of sdLDL involves cholesteryl ester transfer protein (CETP), which facilitates the transfer of cholesterol ester and TAG from VLDL to LDL and hepatic lipase. Since LDL-R has a low affinity for smaller LDL, these particles remain in the circulation. Hyperlipidemia can be increased by low activity of lipoprotein lipase or by a high level of apo C-III, an inhibitor of lipoprotein lipase.

Further, VLDL synthesis and export are impaired in NASH. The blockade of hepatic VLDL secretion results in accumulation of triglycerides in the liver. Microsomal triglyceride transfer protein (MTTP) is essential for formation of VLDL in the liver. Human MTTP polymorphisms lead to decreased MTTP activity and VLDL export and are associated with increased intracellular TAG accumulation.

An abnormal lipoprotein concentration in plasma reflects disturbances in homeostasis of their lipoproteins' major lipid components: TAG, cholesterol, and cholesterol esters. Excessive accumulation of TAG in the liver is the hallmark of NAFLD. The sources of fat that contribute to hepatic steatosis are dietary fats, increased lipolysis of TAG stored in adipose tissue, and de novo synthesis in hepatocytes.

A major cause of TAG accumulation in NAFLD is the liver's inability to regulate lipogenesis in the transition from a fasting to a fed state. Increased lipogenesis in hepatic steatosis may be caused by stimulated synthesis of TAG and decreased ß-oxidation of fatty acid because of decreased production of malonyl-CoA. De novo synthesis is increased in NAFLD and is associated with depletion of n-3 polyunsaturated fatty acids, a consequence of decreased liver Δ^{-6} and Δ^{-5} desaturase in obese NAFLD patients.

Choline deficiency can also lead to fatty liver by impairing VLDL packaging and secretion from hepatocytes. Choline is an essential nutrient required for formation of phosphatidylcholine (PC), which is required for formation of VLDL particles for export of TAG from liver cells. Choline (acting as a lipotropic factor) aids in recovery from fatty liver caused by the inhibition of protein synthesis. Fatty liver caused by choline deficiency is usually enhanced by vitamin E deficiency, as vitamin E, along with selenium, protects against fatty-liver formation by preventing lipid peroxidation. Fatty liver can also result from a deficiency of essential fatty acids, pyridoxine, and pantothenic acid.

Alcoholism

Chronic alcohol consumption is hepatotoxic and leads to TAG accumulation and ultimately to liver cirrhosis. In hepatocytes, the action of alcohol dehydrogenase followed by aldehyde dehydrogenase result in the production of NADH and an increase in cytosolic [NADH]/[NAD$^+$]. This increased production of NADH inhibits ß-oxidation and promotes TAG formation by affecting the ratio of glycerol-phosphate to dihydroxyacetone phosphate in the liver.

$$\text{Ethanol} + \text{NAD}^+ \xrightarrow{\text{Alcohol Dehydrogenase}} \text{Acetaldehyde} + \text{NADH} + \text{H}^+$$

Alcoholism decreases secretion of VLDL, leading to the accumulation of fat (mainly TAG) and hepatic malfunction.

CHOLESTEROL AND CARDIOVASCULAR DISEASE

High plasma cholesterol, particularly as LDL, is strongly linked to increased incidence of coronary heart disease (CHD). In contrast, an increase in HDL in plasma correlates with lowered risk of CHD. Because HDL transports cholesterol from the extrahepatic tissues to the liver for conversion into bile salts, there is decreased cholesterol accumulation in the arteries and, hence, decreased incidence of CHD.

REDUCTION OF PLASMA CHOLESTEROL CONCENTRATION BY CHANGE OF DIET

Some people have a genetic constitution that protects them from high plasma cholesterol concentration. They can consume a diet rich in cholesterol and maintain a cholesterol concentration well below 200 mg/dl. However, most adults do not have this capability and, therefore, have to use dietary or drug treatment (which was described in the context of cholesterol metabolism) to reduce their plasma cholesterol concentration.

Substitution in the diet of polyunsaturated fatty acids for saturated fatty acids is the most beneficial method for reducing plasma cholesterol, as this increases plasma HDL and decreases plasma LDL. It is well established that people who eat diets rich in fish (rich in ω-3 polyunsaturated fatty acids, particularly C22:5 and C22:6) have low plasma cholesterol and a reduced risk of developing heart disease.

ABNORMAL LIPOPROTEINS

Lipoprotein (a) [LP(a)]

Lipoprotein (a) is a complex particle of a variant of human plasma LDL that contains an additional large polymorphic glycoprotein called apolipoprotein (a) that is covalently linked by a disulphide bond to apo B-100 of LDL. Plasma of healthy individuals contains only trace amounts (secreted primarily by the liver) of this lipoprotein. Increased amounts increase the risk of coronary heart disease. The structure of apolipoprotein (a) is related to that of plasminogen, the precursor of the blood protease whose substrate is fibrin. High plasma LP(a) is a primary genetic risk factor for coronary heart disease and stroke.

ß-Very-low-density lipoprotein (ß-VLDL)

These lipoproteins migrate during electrophoresis as LDL, but when separated by ultracentrifugation, they show a density range identical to that of VLDL. They are derived from CMs and are usually termed floating ß-lipoprotein. They are rich in cholesterol and are usually taken up by macrophages. The presence of high levels of ß-VLDL in plasma is associated with increased risk of atherosclerosis.

Lipoprotein-X (LP-X)

This is an abnormal LDL that migrates like LDL during electrophoresis and may be found in the plasma of individuals with intra- or extra-hepatic cholestasis or familial lecithin cholesterol acyltransferase (LCAT) deficiency and in newborn infants with immature liver function. LP-X is a lamellar particle characterized by a high content of phospholipid and unesterified cholesterol. The protein content is dominated by albumin in the core and apo C on the surface. LP-X lacks apolipoprotein B-100 and does not interact with the LDL receptor. It cannot deliver cholesterol to the liver or down-regulate cholesterol synthesis.

DISEASES OF LIPOPROTEIN METABOLISM

A group of inherited disorders arises as a result of abnormal synthesis, processing, and catabolism of plasma lipoprotein particles:

Hypolipoproteinemia

Abetalipoproteinemia

This was described in chapter 1.

Familial hypobetalipoproteinemia

This is characterized by plasma LDL concentration lower than 50% of normal accompanied by normal chylomicron formation.

Tangier disease (familial alpha-lipoprotein deficiency)

This rare autosomal-recessive disease results from a severe deficiency or absence of plasma HDL and leads to hepatomegaly and splenomegaly. It is characterized by accumulation of cholesterol esters in many tissues, including that of peripheral nerves, liver, spleen, tonsils, lymph nodes, bone marrow, and cornea. Plasma apo A-I is extremely low, plasma cholesterol and phospholipids are reduced, and plasma TAG is normal or increased.

Hyperlipoproteinemia (hyperlipidemia)

Hyperlipoproteinemia may be primary or secondary. The former is usually genetically determined, and the latter is associated with disorders such as nephrotic syndrome and hypothyroidism. Hyperlipoproteinemia has been classified into five types (I–V) according to the changes in plasma lipoprotein pattern, by Fredrickson (1965).

Hyperlipidemia type I

This arises from accumulation of CMs caused by apo C-II deficiency or lipoprotein lipase deficiency. Both forms are characterized by a very high concentration of TAG in plasma due to very slow clearance of CMs from the circulation and are induced by the ingestion of fat. Therefore, decreasing the quantity of fat and increasing the proportion of carbohydrates in the diet may reduce the symptoms.

Hyperlipidemia type II (familial hypercholesterolemia)

This arises from genetic defects in the synthesis, processing, or function of the LDL receptor. Two types (IIa and IIb) are recognized. Both are characterized by elevated levels of plasma LDL, but in type IIb, VLDL is also elevated. Thus, individuals with type IIa have elevated plasma cholesterol, and those with type IIb show a rise in cholesterol and TAG. Type IIb is often associated with type 2 diabetes mellitus, metabolic syndrome, and chronic renal disease and is one of the most common disorders of hyperlipidemia. Individuals with this type are at high risk for cardiovascular disease (CVD).

Homozygous patients may suffer myocardial infarction before age 20 due to a very high level of plasma cholesterol.

Note that hypercholesterolemia may also be produced by lysosomal cholesterol ester hydrolase deficiency (Wolman's disease or cholesterol-ester storage disease).

Hyperlipidemia type III

This is characterized by the occurrence of a ß-migrating VLDL (ß-VLDL) fraction that consists of cholesterol-enriched remnants of CM and VLDL. Plasma concentration of cholesterol and TAG are elevated. It arises from an abnormality in apo E, which is required for the liver's uptake of VLDL and CM particles. The disease usually leads to xanthomas (yellow deposits in the skin) and atherosclerosis. Weight reduction and diets that are low in saturated fatty acids and cholesterol and that contain complex carbohydrates may reduce the symptoms.

Hyperlipidemia type IV

This is the most common type and is characterized by a high concentration of endogenously formed TAG (VLDL). Plasma cholesterol rises in proportion to the very high levels of TAG, but plasma levels of both HDL and LDL are subnormal. The molecular defect in familial forms is unknown. This hyperlipidemia pattern is frequent in patients with obesity, alcoholism, or diabetes mellitus. Treatment by weight reduction and diets that contain complex carbohydrates, polyunsaturated fatty acids, and small amounts of cholesterol along with hypolipidemic drugs is usual.

Hyperlipidemia type V

This is characterized by elevated plasma concentration of CMs and VLDL, leading to increased plasma concentration of TAG and cholesterol and to xanthomas. Weight reduction followed by a moderate intake of carbohydrates and fat in the diet is the usual treatment.

Familial lecithin-cholesterol acyltransferase (LCAT) deficiency

Familial LCAT deficiency is characterized by the absence or near absence of LCAT activity in plasma. It leads to corneal opacities, anemia, proteinuria, and renal failure. Individuals with familial LCAT deficiency show HDL deficiency, which leads to blocked reverse-cholesterol transport. HDL remains in nascent discs in stacks incapable of taking up cholesterol and of converting it to cholesterol ester. This is accompanied by alterations in plasma VLDL and LDL levels.

Fish-eye disease, another form of LCAT deficiency (partial LCAT deficiency), is characterized by absence of LCAT activity towards HDL but presence of LCAT activity towards LDL. This type of LCAT deficiency leads only to corneal opacities and is characterized by low HDL cholesterol, elevated TAG, and multiple plasma-lipoprotein abnormalities.

CONCLUSION

- Plasma lipoproteins are complexes of a variety of lipids with one or more specific proteins called apolipoproteins.
- They include chylomicron, high-density lipoprotein, intermediate-density lipoprotein, low-density lipoprotein, and very-low-density lipoprotein.
- They undergo specific changes in their lipid and protein content before being taken up by body tissues via specific receptors.
- Inherited defects in secretion, release, catabolism, or uptake of a specific lipoprotein can lead to clinically important hyper- or hypolipoproteinemias.

SELECTED READINGS

Brewer, H. B., Gregg, R. E., Hoeg, J. M. et al. (1988). Apolipoproteins and lipoproteins in human plasma: An overview. *Clinical Chemistry* 34: B4–B8.

Brown, M. S., and Goldstein, J. L. (1986). A receptor mediated pathway for cholesterol homeostasis. *Science* 232:34–47

Davis, R. A. (1991). Lipoprotein structure and secretion. In D. E. Vance and J. E. Vance (Eds.), *Biochemistry of lipids, lipoproteins, and membranes.* New York: Elsevier, 403–426.

Fielding C. J., and Fielding P. E: (1985). Metabolism of cholesterol and lipoproteins. In D. E. Vance and J. E. Vance (Eds.), In: *Biochemistry of Lipids and membranes.* Menlo Park, CA: Benjamin/Cummings, 404–474.

Fon Tacer, K., and Rozman, D. (2011). Nonalcoholic fatty liver disease: Focus on lipoprotein and lipid deregulation. *Journal of Lipids* http://dx.doi.org/10.1155/2011/783976.

Franceschini, G., Busnach, G., Chiesa, G., et al. (1991). Management of lipoprotein-X accumulation in severe cholestasis by semi-selective LDL-apheresis. *American Journal of Medicine* 90: 633–638.

Fredrickson, D.S. and Lee, R.S. A. (1965). System for phenotyping hyperlpoproteinemia. Circulation 31: 321-327.

Havel, R. J., and Kane, J. P. (1995). Introduction: Structure and metabolism of plasma lipoproteins. In C. R. Scriver, A. L. Beaudet, W. S. Sly, et al. (Eds.), *The metabolic and molecular bases of inherited disease,* 7th ed. New York: McGraw-Hill, 1841–1851.

Kane, J. B., and Havel, R. J. (1986). Treatment of hypercholesterolemia. *Annual Review of Medicine* 37: 427–435.

Lewis, G. F., and Rader, D. J. (2005). New insights into the regulation of HDL metabolism and reverse cholesterol transport. *Circulation Research* 96: 1221–1232.

Lieber, C. S. (1988). Biochemical and molecular basis of alcohol induced injury to liver and other tissues. *New England Journal of Medicine* 319:1639–1650.

Mahley, R. W., Innerarity, T. L., Rall, S. C. Jr., et al. (1984). Plasma lipoproteins: Apolipoprotein structure and function. *Journal of Lipid Research* 12: 1277–1294.

Sörös, P., Bötter, J., Maschek, H., et al. (1998). Lipoprotein-X in patients with cirrhosis: Its relationship to cholestasis and hypercholesterolemia. *Hepatology* 28 (5):1199–1205.

Third Report of the National Cholesterol Education Program (NCEP), Expert Panel on Detection, Evaluation, and Treatment of High Blood Cholesterol in Adults (Adult Treatment Panel III). (2002). Final report. *Circulation* 106: 3143–3421.

Williams, D. L. (1985). Molecular biology in arteriosclerosis research. *Arteriosclerosis* 5: 213–227.

Chapter 9: Metabolism of Steroid Hormones

Introduction

All steroid hormones are synthesized from cholesterol. They include glucocorticoids (cortisol and corticosterone), mineralocorticoids (e.g. aldosterone), and the sex hormones (androgens, estrogens, and progestins). The structure of some important steroid hormones was presented in chapter 1. Synthesis of steroid hormones takes place in the adrenal cortex (cortisol and aldosterone), ovaries and ovarian corpus luteum (estrogens, progestins), and testes (testosterone). These hormones are transported in the blood from their sites of synthesis to their target organs by specific plasma steroid–carrier proteins and, to a minor extent, by albumin. Activation of their biosynthesis involves an increase in cholesterol ester hydrolysis and cellular uptake of cholesterol into the mitochondria of cells in the appropriate organ. These hormones act by binding to specific intracellular receptors that enter the nucleus of the target cells and trigger the expression of hormone-sensitive genes. Several metabolic diseases are associated with abnormal steroid hormone metabolism.

Major Functions of Steroid Hormones

- Aldosterone stimulates reabsorption of Na^+ and excretion of K^+ and H^+ by renal tubules, thus leading to an increase in blood volume and pressure.
- Cortisol promotes gluconeogenesis, suppresses the inflammation reaction, and stimulates breakdown of fat and protein.
- Estrogens control the menstrual cycle and promote the development of female secondary sex characteristics.
- Progesterone is responsible for the secretory phase of the menstrual cycle and mammary glands. It also stimulates nidation and maturation of the fertilized ovum.
- Testosterone stimulates spermatogenesis and promotes development of male secondary sex characteristics.

Biosynthesis of Steroid Hormones

Steroid hormone synthesis from cholesterol begins when the anterior pituitary hormone corticotropin (ACTH) stimulates synthesis of mineralocorticoids and glucocorticoids in the adrenal cortex, and when follicle-stimulating hormone (FSH) and luteinizing hormone (LH) stimulate the synthesis of gonadal steroid hormones.

The initial and rate-limiting step in steroid-hormone synthesis is conversion of cholesterol to the 21-carbon pregnenolone by cholesterol desmolase, also known as P450 cholesterol side-chain cleavage enzyme (P450scc), which hydroxylates the side chain of cholesterol at C20 and C22 and cleaves it to produce pregnenolone and isocaproaldehyde (figure 9.1). This requires molecular O_2, NADPH, and a cytochrome P450.

BIOSYNTHESIS OF ADRENAL CORTICAL STEROID HORMONES

The pathways for conversion of cholesterol to adrenocortical steroid hormones are shown in figure 9.1. Mineralocorticoid synthesis occurs in the adrenal zona glomerulosa and begins with conversion of pregnenolone to progesterone by 3-hydroxysteroid dehydrogenase and $\Delta^{5.4}$ isomerase. The former converts the 3-OH group of pregnenolone to a 3-keto group, and the latter moves the double bond from the B ring to the A ring to form progesterone. Hydroxylation of progesterone at C21 then produces 11-deoxycorticosterone (DOC), which is active in Na^+-retention. The DOC is further hydroxylated at C11 to produce corticosterone, which has glucocorticoid and weak mineralocorticoid activity. Finally, corticosterone is converted to aldosterone by mitochondrial 18-hydroxylase. Production of aldosterone is induced by a low $Na^+{:}K^+$ ratio in plasma and by the hormone angiotensin II, which exerts its effects through the phosphatidylinositol-4,5-bisphosphate (PIP2) pathway to form inositol trisphosphate and diacylglycerol and to activate protein kinase C (PKC).

Glucocorticoids are synthesized from progesterone in the adrenal zona fasciculata through action of 17α-hydroxylase and 21 α -hydroxylase in the endoplasmic reticulum to form 17 α -hydroxyprogesterone and 21α -hydroxyprogesterone respectively. Finally, in the Mitochondria, cortisol is formed from 21 α -hydroxyprogesterone by11ß-hydroxylase.

Androgen synthesis occurs in the adrenal zona reticularis from progesterone, which is converted to dehydroepiandrosterone (DHEA) by the action of 17 α -hydroxylase followed by side-chain cleavage.

BIOSYNTHESIS OF SEX HORMONES

Progesterone is formed as described above for adrenal steroids. Biosynthesis of androgens and estrogens also requires the production of pregnenolone from cholesterol as described above for the adrenal steroids, but for these sex hormones, this process is stimulated by luteinizing hormone (LH).

Luteinizing hormone (LH) acting on Leydig cells of the testes stimulates cAMP production and production of testosterone de novo from cholesterol. The FSH is produced by the anterior pituitary and targets its receptor on the cell surface of granulosa cells of the ovaries and the Sertoli cells of the testes. In the ovaries, FSH induces an aromatase, which converts androgens to estrogens and induces follicular maturation. In addition, the granulosa cells produce the hormone inhibin B, which exerts negative feedback upon FSH. In the testes, FSH upregulates Sertoli cell action and plays a crucial role in spermatogenesis.

Male sex hormones (androgens)

Figure 9.2 shows the biosynthesis of androgens in testes. Testosterone is formed from progesterone by 17α-hydroxylase, C17-20 lyase, and 17ß-hydroxysteroid dehydrogenase. Testosterone is converted to dihydrotestosterone in some androgen target cells by 5 α -reductase in the endoplasmic reticulum and nucleus. Testosterone is also formed from pregnenolone by the dehydroepiandrosterone pathway (see figure 9.2). This involves17 α -hydroxylase, C17-20 lyase, 17 ß -hydroxysteroid dehydrogenase, 3 ß –hydroxysteroid dehydrogenase, and $\Delta^{5,4}$ isomerase. The testes also produce 17 ß -estradiol (a female sex hormone) in small amounts, but most of this is formed by aromatization of androstenedione and testosterone in peripheral tissues.

Female sex hormones (estrogens, progestins)

The naturally occurring estrogens in women are estrone (E1), estradiol (E2), and estriol (E3). They are formed by the ovaries and placenta during pregnancy upon stimulation by FSH. The reactions involve formation of progesterone, androstenedione, and testosterone as described above for androgen formation. Androstenedione is converted into testosterone by 17ß-hydroxysteroid dehydrogenase, whereas conversion of androstenedione and testosterone into estrone and 17 ß -estradiol, respectively, requires aromatase in the presence of O_2 and NADPH. Figure 9.3 shows the biosynthesis of female sex hormones.

The primary source of estrogens in postmenopausal women is the conversion of androstenedione to estrone. In pregnancy, about 50% of the 17 ß -estradiol is derived from aromatization of adrenal androgen.

TRANSPORT OF STEROID HORMONES

Steroid hormones are transported in plasma bound to corticosteroid-binding globulin, sex hormone-binding protein, androgen-binding protein, or albumin. In addition to this transport role, these proteins also protect the steroids from metabolism and inactivation, provide free steroids for entry into their target cells, and assist in maintaining the level of steroid hormones in the blood.

Corticosteroid-binding globulin (CBG)

Corticosteroid-binding globulin, also called transcortin, transports glucocorticoids and progesterone. This plasma ß -globulin binds about 80% of the total 17-hydroxysteroids in the circulation. About 75% of the cortisol is bound to transcortin and 22% to albumin. Only about 10% of the total aldosterone in blood is bound to transcortin and about 60% is bound to albumin.

Sex-hormone-binding globulin (SHBG)

SHBG binds more tightly to androgens than to albumin. In the circulation, 1 to 3% of testosterone is free, 10% is bound to SHBG, and the remainder is bound to albumin. Only an unbound steroid hormone can permeate cells and bind to intracellular receptors.

Androgen-binding protein (ABP)

Androgen-binding protein is a glycoprotein that binds specifically to testosterone, 17-ß-estradiol, and dihydrotestosterone. Testosterone and FSH stimulate Sertoli cells in the testes to produce ABP, which is required to maintain a high local concentration of testosterone in the vicinity of the developing germ cells within the seminiferous tubules.

Figure 9.1

Biosynthetic pathways of adrenal-cortical steroid hormones.

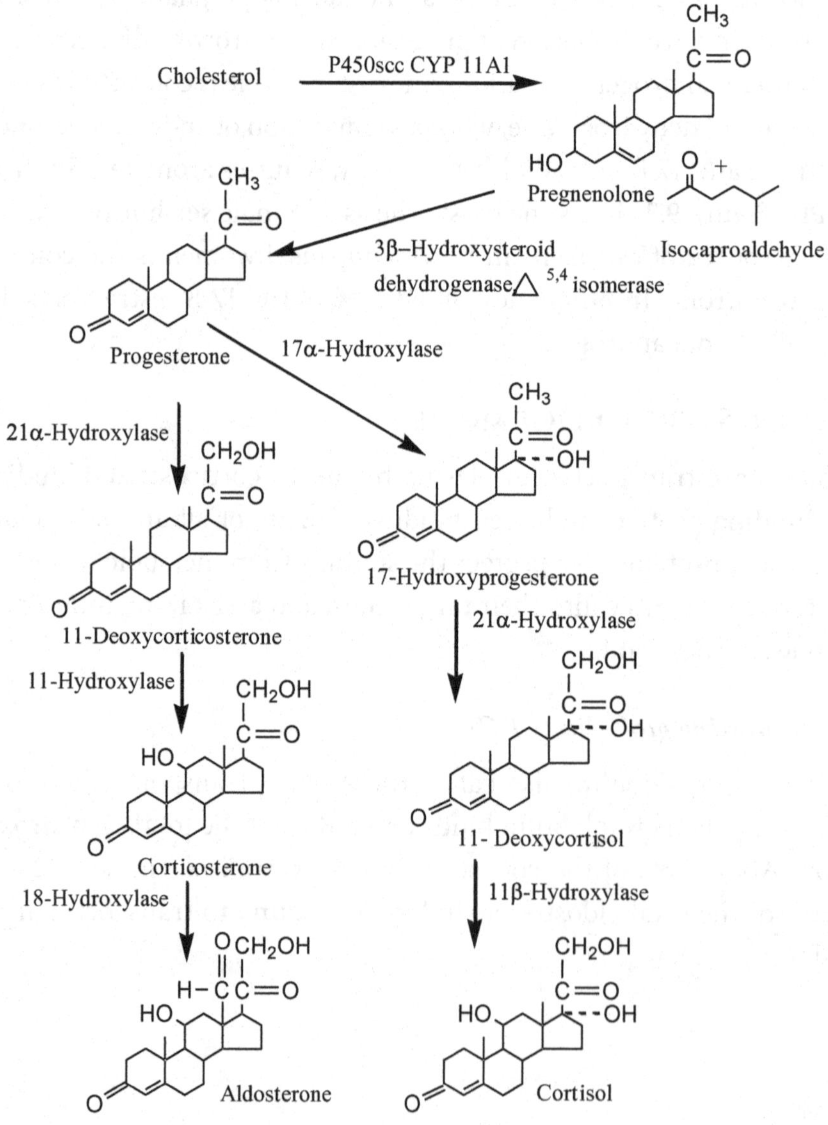

Figure 9.2

Conversion of cholesterol to male sex hormones.

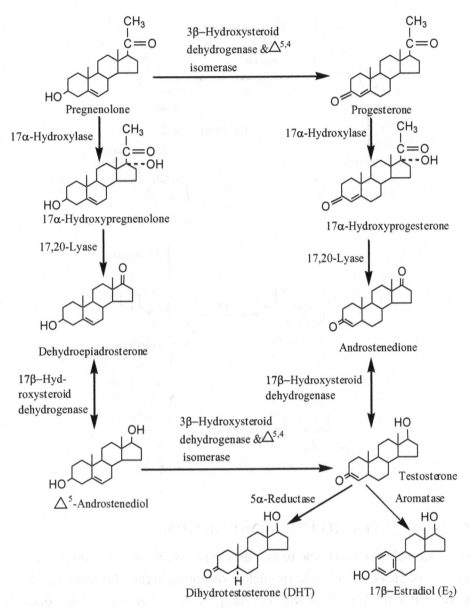

Figure 9.3

Conversion of cholesterol to female sex hormones.

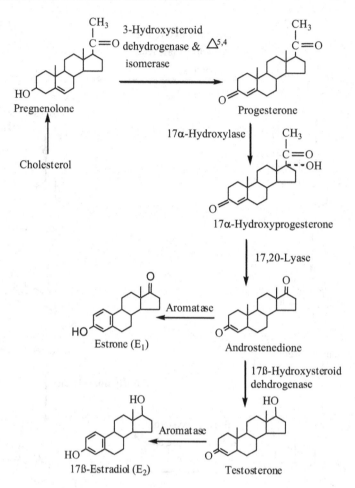

MECHANISM OF STEROID HORMONE ACTION

Whereas non-steroid hormones bind to membrane receptors on their target cells, steroid hormones interact with specific cytosolic or nuclear receptors in their target cells. Glucocorticoid receptors are located in the cytosol, whereas receptors for other steroid hormones are located within the nucleus. The receptor-hormone complex in the nucleus binds to regulatory DNA sequences (hormone-responsive element, HRE) of their target genes and cause activation or inhibition of the specific genes. Increased or decreased expression of these genes alters cell function. Figure 9.4 shows a model for the mechanism of action of steroid hormones. The numbers in the sequence below correspond to the numbers in figure 9.4.

1. Release of steroid hormone from the transport protein in circulation.
2. Free hormone enters by diffusion into the cytosol or the nucleus of the target cell.
3. Steroid hormone binds to cytosolic or nuclear receptor.
4. Steroid hormone exposes the receptor's DNA binding.

5. The binding of steroid hormone to the inactivated receptor causes release of heat-shock proteins from the receptor, activating it. (When attached to the cytosol receptor, the heat-scock proteins block the DNA-binding domain.)

6. Cytosolic hormone-receptor complex is translocated into the nucleus after the DNA-binding domain is activated and exposed.

7. Binding of activated hormone-receptor complex to specific regulatory DNA sequences (hormone-responsive elements, HRE) on hormone-sensitive genes leads to activation or inhibition of the activity of specific genes.

8. Binding of hormone-receptor complex to HRE activates transcription. New mRNAs are translocated to the cytosol and assembled into translation complexes for the biosynthesis of new proteins.

9. Newly synthesized proteins alter the metabolism and function of the target tissue cells. This effect can last for relatively long periods (hours to days).

EXCRETION OF STEROID HORMONES

The liver converts steroid hormones to excretion products by making them more water soluble through hydroxylation reactions, reduction of the saturated bonds of the steroid structure, and conjugation with glucuronic acid or sulfate. Most of the conjugated products (approximately 70%) are excreted in the urine, and the remainder are secreted into the bile and finally into the feces.

ABNORMALITIES IN STEROID HORMONE SECRETION

Hypoadrenocorticism

This may be caused by an intrinsic defect of the adrenal cortex (in primary hypoadrenocorticism) or corticotrophin-releasing hormone (CRH) or ACTH deficiency (in secondary hypoadrenocorticism) or by a lack of responsiveness to an adrenocortical hormone.

Congenital adrenal hyperplasia (CAH)

CAH is a type of hypoadrenocorticism caused by the deficiency of an enzyme involved in steroid-hormone synthesis because of a congenital genetic abnormality. This leads to accumulation of the substrates of the affected enzyme, deficiency of the end products, and increased production of steroid hormones by alternative pathways. The various forms of CAH are inherited in an autosomal-recessive manner. Hyperplasia of the adrenal cortex results from excessive secretion of ACTH. Table 9.1 presents a summary of CAH.

Addison's disease (primary adrenal insufficiency)

This is characterized by hypoglycemia, extreme sensitivity to insulin, intolerance to stress, weight loss, anorexia, nausea, severe weakness, and pigmentation of skin and mucous membranes. Low blood pressure, low glomerular filtration rate, and decreased ability to excrete water are usually also present. Patients have high plasma K$^+$, low Na$^+$ levels, and high counts of eosinophils and lymphocytes in the blood. The disease results from glucocorticoid insufficiency as a result of adrenocortical atrophy or destruction by tuberculosis or tumor.

Secondary adrenal insufficiency is caused by deficiency as a result of infection or a tumor of the pituitary gland. The metabolic effects are similar to those described above for the primary condition, but there is no pigmentation.

Cushing's syndrome

Excess glucocorticoid production commonly results from abnormal secretion of ACTH by the anterior pituitary, adrenal cortical tumor, or ectopic production of ACTH by a neoplasm. The disease is characterized by obesity, elevated blood glucose, and hypertension. Patients develop a characteristic puffy moon face and a 'buffalo hump' from fat deposits, the latter between the shoulders. A similar condition may result from pharmacologic use of steroids.

Androgen insensitivity syndrome (AIS)

This condition (also called testicular feminization) is an X-linked disease usually present with a 46-XY karyotype, incompletely descended testes, and female or partially masculinized external genitalia caused by loss-of-function mutations in the androgen receptor (AR) gene, which results in tissue resistance to androgen. There are three clinical subgroups based on the genital phenotype: complete (CAIS), partial (PAIS), and mild or minimal (MAIS).

Affected individuals can have normal female external genitalia but may have a short blind-ending vagina; be missing Wolffian duct-derived structures such as epididymides, vas deferens, and seminal vesicles; or be missing a prostate gland. Breast development occurs but pubic and axillary hair are absent at puberty, so such people born with unambiguously female anatomy will not be suspected of having this condition until the onset of puberty.

5α-reductase-2 deficiency

This is a disorder of sexual development (DSD) transmitted as an autosomal-recessive condition in which 46-XY subjects with bilateral testes and normal testosterone formation have impaired virilization during embryogenesis due to defective conversion of testosterone to dihydrotestosterone (DHT). This conversion requires the membrane-bound steroid 5α-reductase, which exists in two isoforms. Type 1 isoenzyme is expressed at low levels in the prostate, whereas type 2 isoenzyme is expressed at high levels in the prostate and in many other androgen-sensitive tissues. The deficiency is caused by mutations in the gene that encodes the steroid 5α-reductase-2 isoenzyme.

Male hypogonadism

This is characterized by impaired testicular function, which can affect spermatogenesis, testosterone synthesis, or both. It can result from a primary testicular disorder or be secondary to hypothalamic-pituitary dysfunction. Hypergonadotropic hypogonadism is characterized by decreased testosterone production and elevated levels of follicle-stimulating hormone and luteinizing hormone.

Hypogonadotropic hypogonadism is a consequence of a congenital or acquired disease that affects the hypothalamus, the pituitary gland, or both so that secretion of gonadotropin-releasing hormone (GnRH) is absent or inadequate, leading to absent or inadequate biosynthesis of pituitary gonadotropins.

Klinefelter syndrome affects males and arises from duplication of the X chromosome so that the karyotype is 47-XXY. Those with this condition experience testicular dysgenesis with decreased production of testosterone and compensatory increase of FSH and LH, along with infertility and swelling of the breasts.

Female hypogonadism

One form of this condition is Turner syndrome (monosomy X), which affects females and arises from the lack of an X chromosome so that the karyotype is 45-XO. Those with this condition have decreased production of estrogen and progesterone and a compensatory increase in FSH and LH, lack the appearance of puberty, and experience amenorrhea.

Figure 9.4

Mechanism of action of steroid hormones.

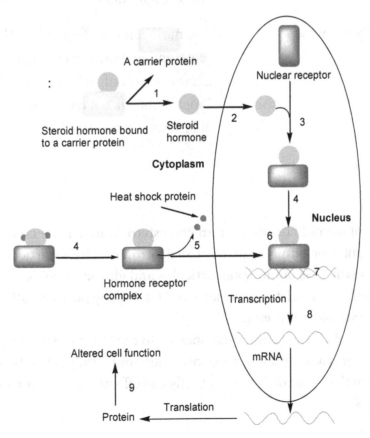

Table 9.1

Summary of congenital adrenal hyperplasias.

Enzyme deficiency	Common features
Cytochrome P450 scc	Rare. There is no steroid hormone production and cholesterol accumulation in adrenocortical cells
11-β-Hydroxylase β-hydroxylase	Representing approximately 5% of CAH cases. Low serum levels of aldosterone, cortisol, & corticosterone. Increased production of deoxycorticosterone leads to fluid retention and hypertension.
3-β-Hydroxysteroid Dehydrogenase	Rare. Glucocorticoids, mineralocorticoids, androgens and estrogens formation is blocked.
17- α -Hydroxylase	Rare. Block in production of cortisol and sex hormones.
21-α-Hydroxylase	The most common form of CAH, representing more than 90% of cases. Elevated level of ACTH. Progesterone is not converted to aldosterone and cortisol.

CONCLUSION

- Synthesis of steroid hormones from cholesterol is under the control of hormones from the anterior pituitary gland. Corticotropin (adrenocorticotropic hormone) stimulates synthesis of mineralocorticoids and glucocorticoids in the adrenal cortex, and follicle stimulating hormone and luteinizing hormone stimulate synthesis of gonadal steroid hormones.
- Secreted steroid hormones are transported in the blood by specific plasma steroid–carrier proteins and, to a lesser extent, by albumin. They act by binding to specific receptors in the nuclei of their target cells and effecting expression of hormone-sensitive genes.

- Mineralocorticoids increase blood volume and pressure, whereas glucocorticoids promote gluconeogenesis, suppress the inflammation reaction, and stimulate breakdown of fat and protein.
- Estrogens control the menstrual cycle and promote the development of female secondary sex characteristics.
- Progesterone is responsible for the secretory phase of the menstrual cycle and for development of mammary glands. It also stimulates nidation and maturation of the fertilized ovum.
- Testosterone stimulates spermatogenesis and promotes development of male secondary sex characteristics.
- Several metabolic diseases are associated with abnormal steroid hormone metabolism.

SELECTED READINGS

Evans, R. M. (1988). The steroid and thyroid hormone receptor superfamily. *Science* 240 (4854): 889–895.

Galani, A., Kitsio-Tzeli, S., Sofokleous, C., et al. (2008). Androgen insensitivity syndrome: Clinical features and molecular defects. *Hormones* 7 (3):217–229.

Miller, W. L., and Auchus, R. J. (2011). The molecular biology, biochemistry, and physiology of human steroidogenesis and its disorders. *Endocine Reviews* 32 (1): 81–151.

O'Malley, B. W. (1984). Steroid hormone action in eucaryotic cells. *The Journal of Clinical Investigations* 74 (2):307–312.

Schwabe, J. W. R., and Rhodes, D. (1991). Beyond zinc fingers: Steroid hormone receptors have a novel structural motif for DNA recognition. *Trends in Biochemical Scinces* 16: 291–296.

Whali, W. and Martinez, E. (1991). Superfamily of steroid nuclear receptors: Positive and negative regulators of gene expression. *FASEB. Joural* 5: 2243–2249.

Chapter 10: Fat-Soluble Vitamins

Introduction

The fat-soluble vitamins A, D, E, and K are isoprene derivatives. They are essential dietary constituents, because the body cannot synthesize them in sufficient quantities. Since they are hydrophobic molecules, they must be absorbed and transported like other lipids taken in through the diet. Vitamin A is required for vision, vitamin D for calcium and phosphorus metabolism, vitamin E as an antioxidant, and vitamin K for blood coagulation. Inadequate dietary intake of vitamin A results in night blindness and xerophathalmia, of vitamin D in rickets (in young children) and osteomalacia (in adults), of vitamin E in neuromuscular and neurological problems and anemia, and of vitamin K in bleeding and hemorrhage. Excessive intake of vitamin A or D can result in toxicity.

Vitamin A (Retinol)

Retinol is a polyisoprenoid compound that contains a cyclohexenyl ring. Several compounds derived from animal sources exhibit the biologic activity of vitamin A and are referred to as vitamin A. They include retinol (vitamin A), retinal, and retinoic acid (see figure 10.1). The term retinoid has been used to describe natural or synthetic forms of vitamin A, which may or may not exhibit the biologic activity of vitamin A.

Provitamin A (ß-carotene)

Provitamin A, or ß-carotene (an isoprenoid compound), is an important precursor of vitamin A. It occurs widely in vegetables, including carrots. It consists of the equivalent of 2 molecules of retinol joined at the end of their carbon chains, and it is cleaved to the vitamin in the intestinal mucosal cells (figure 10.1). Provitamin A is not efficiently metabolized to vitamin A so that 6 µg of ß-carotene is required to produce 1 µg of vitamin A. Consequently, the dietary requirement is expressed as retinol equivalents (RE). Two forms of vitamin A are available in the human diet: preformed vitamin A (retinol and its esterified form, retinyl ester) and provitamin A carotenoids, (ß-carotene, α-carotene and ß-cryptoxanthin). The RDA for adult females and males is 800 and 1000 RE, respectively.

Digestion and absorption of vitamin A

Dietary retinol esters are emulsified with other dietary lipids in the presence of bile salts and hydrolyzed in the intestinal mucosa to release free fatty acids, and then they are cleaved by ß-carotene deoxygenase in the presence of oxygen to 2 molecules of retinal (retinaldehyde, see figure 10.1). This is then reduced in the intestinal mucosa to retinol by retinaldehyde reductase in presence of NADPH. A small amount of retinal is oxidized to retinoic acid. Retinol derived from the hydrolysis of dietary retinol esters or from the cleavage of carotene is esterified to saturated long-chain fatty acid in the intestinal mucosa and secreted as a component of chylomicrons into the lymphatic system. These then enter the blood stream and are converted to chylomicron remnants, which are taken up by hepatocytes and stored.

Storage, release, and transport

The liver contains supplies of retinol esters such that it can be released to systemic circulation for several months. Retinol is transported to the extrahepatic tissues by plasma retinol-binding protein (RBP). Once inside the cells, retinol is transported by a cellular retinol-binding protein to the nucleus, where retinal acts similarly to steroid hormones. Retinoic acid is transported in blood bound to albumin, and once inside cells, it also acts like a steroid hormone.

Figure 10.1

Conversion of ß-carotene to retinal, retinol, and retinoic acid.

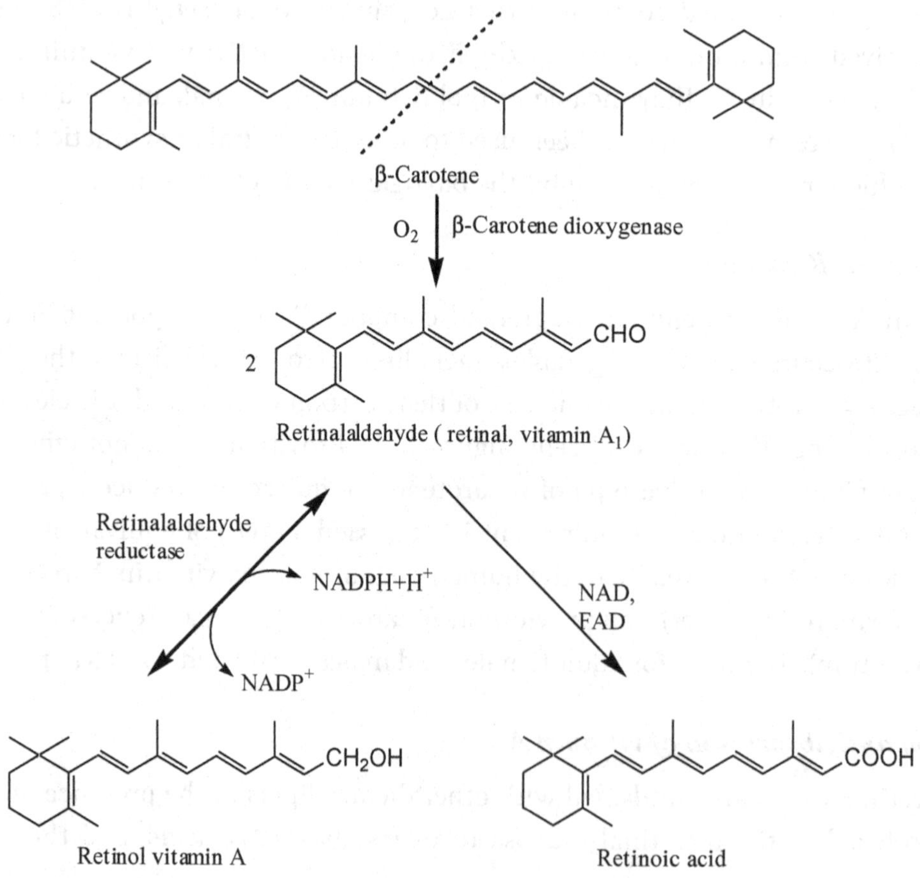

Functions of vitamin A

Visual cycle

Retinal is a component of the visual pigment in the rod cells of the retina which are responsible for vision in poor light. Rhodopsin consists of 11-*cis*-retinal (an isomer of all-*trans*-retinal) specifically bound to a lysine residue of the protein opsin by a Schiff base linkage. Absorption of a photon of radiation by 11-*cis*-retinal triggers a series of photochemical reactions that result in bleaching of the visual pigment and the release of all -trans-retinal and opsin (figure 10.2). The result of these processes triggers a nerve impulse that is transmitted to the brain. The numbers in the sequence below correspond to numbers in figure 10.2.

1. The compound 11-*cis*-retinal is present in the unstimulated photoreceptor as a component of rhodopsin.
2. The absorption of a photon of radiation by 11-*cis*-retinal leads retinal to isomerize to all-*trans*-retinal and produces the intermediate bathorhodopsin.
3. The Schiff base in bathorhodopsin is deprotonated to produce metarhodopsin II, which is the key to the transduction of photon reception to neural action potential. Metarhodopsin II accomplishes this by activating a cascade that leads to a rapid decrease in cGMP level, closing Na^+ channels in the membrane of the visual receptor (see figure 10.3). The result of this excitation is a pulse of hyperpolarization (a negative signal) in the light receptor cell.
4. After about one second, metarhodopsin II dissociates into all-*trans*-retinal and opsin.
5. Finally, the all-*trans*-retinal is isomerized back to 11-*cis*-retinal, which recombines with opsin to re-form rhodopsin.

Reproduction

All-*trans* retinoic acid (RA), through several intracellular receptors, supports both male and female reproduction and acts as a morphogen in embryonic development. Vitamin A is essential for the maintenance of the male genital tract and spermatogenesis. Vitamin A is also participates in a signalling mechanism to initiate meiosis in the female gonad during embryogenesis, and in the male gonad postnatally.

Figure 10.2

Chemical changes in photoreception.

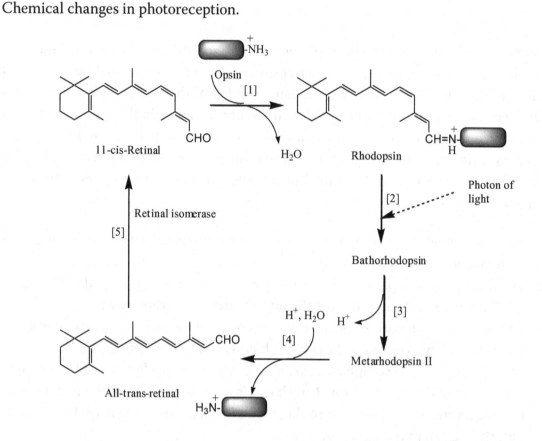

Figure 10.3

Role of metarhodopsin II in the visual cascade. Metarhodopsin II catalyzes the exchange of GTP for GDP on transducin (a trimeric G protein). In this process, two of the transducin subunits (Tβ and Tγ) are released, leaving GTP bound to the α subunit (Tα), which activates phosphodiesterase by removing the inhibitory subunit (I) from it. The activated phosphodiesterase hydrolyzes cGMP to 5GMP. This leads to a rapid decrease of cGMP with subsequent blockage of Na^+ channels in the membrane of the visual receptor. The GTP on Tα is hydrolyzed to GDP, and the inhibitory subunit (I) returns to the phosphodiesterase. The system returns to the resting state by reassembly of the three transducin subunits.

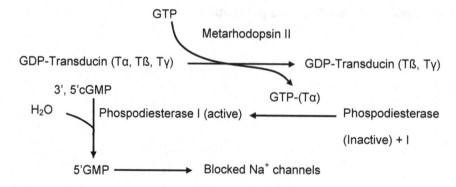

Dietary sources of vitamin A

Carotenes are present in yellow and dark-green vegetables and fruits. Good dietary sources of preformed vitamin A include liver, kidney, egg yolk, cream, and butter.

Vitamin A toxicity (hypervitaminosis A)

Hypervitaminosis A is caused by excessive intake of vitamin A. It is characterized by headache, loss of appetite, hepatosplenomegaly, and dermatitis. Consumption of more than 7.5 mg of retinol per day exposes the cell to unbound retinol because the capacity of RBP to bind retinol is exceeded. Ingestion of excessive quantities of vitamin A should be avoided particularly during pregnancy to protect the developing fetus from congenital malformation as a result.

Vitamin A deficiency

Diets deficient in vitamin A lead to night blindness (defective dark adaptation), usually the first symptom to occur when liver stores are nearly exhausted. If not treated this leads to keratinization of epithelial, lung, genitourinary, and gastrointestinal tract tissues and decreased mucus secretion. Xerophthalmia, dryness of the conjunctiva and cornea, may lead to blindness if not treated. Deficiency may also lead to abnormalities of reproduction, including degeneration of the testes, miscarriage, or the production of malformed offspring. Some of these symptoms respond to administration of vitamin A in the form of retinol or retinyl esters.

VITAMIN D

The term provitamin D refers to the sterols 7-dehydrocholesterol and ergosterol, present in animals and plants, respectively. They are converted in the body to a hormone known as calcitriol, which plays an important role in regulating calcium and phosphorus metabolism. Deficiency of this vitamin causes a net demineralization of bone that leads to rickets in children and osteomalacia in adults.

Formation of vitamin D

Ergosterol differs from 7-dehydrocholesterol only in its side chain, which is unsaturated and contains an extra methyl group. Ergosterol (provitamin D_2) and 7-dehydrocholesterol (provitamin D_3) are converted to ergocalciferol (vitamin D_2) in plants and cholecalciferol (vitamin D_3) in the skin of mammals by exposure to ultraviolet radiation in sunlight (figure 10.4).

Formation of 1,25-dihydroxycholecalciferol (calcitriol)

In the small intestine, dietary vitamin D is absorbed as a component of mixed micelles. In the circulation, vitamin D (from the diet or synthesized in the skin) is transported by vitamin D–binding protein and taken up by the liver, where it is hydroxylated on the 25 position to produce 25-hydroxycholecalciferol, the predominant form of vitamin D in the blood and the primary storage form in liver. This hydroxylation is catalyzed by a microsomal 25-hydroxylase (cytochrome P450-dependent) in the presence of O_2 and NADPH. The 25-hydroxycholecalciferol is then transported by the carrier protein to the proximal tubules of the kidney, where it is hydroxylated in the 1 position to produce 1,25-dihydroxycholecalciferol (calcitriol), the active form of the vitamin. This reaction is catalyzed by microsomal 1α-hydroxylase, which requires cytochrome P450, oxygen, and NADPH. The key enzyme in the synthesis of calcitriol, 1α-hydroxylase, is activated by low plasma phosphate and by parathyroid hormone in response to a low plasma calcium level. It is inhibited by the product of the reaction (through feedback inhibition). When calcitriol increases in plasma, 1-α-hydroxylase enzyme is inhibited and 24-hydroxylase is stimulated to convert 25-hydroxycholecalciferol into 24,25-dihydroxycholecalciferol, which is believed to be inactive. Although the kidney is the primary site of production of both 1,25-dihydroxyvitamin D_3 and 24,25-dihydroxyvitamin D_3, they are also produced by other tissues. For example, the 1-α-hydroxylase and the 24-hydroxylase reactions also occur in bone and placental tissue. Figure 10.4 shows the formation of these two metabolites.

Biological functions of vitamin D

The principal function of calcitriol is to maintain adequate plasma levels of calcium and phosphate. It performs this by stimulating the intestinal and renal absorption of calcium and phosphate and their mobilization from bone.

Calcitriol binds to specific nuclear receptors in the intestine, bone, and kidney and transcriptionally activates genes that encode calcium-binding proteins involved in absorption and reabsorption of both calcium and phosphate. The mechanism of action of calcitriol is similar to that described for steroid hormones (see chapter 9).

Deficiency of vitamin D

Deficiency of vitamin D leads to rickets in children and osteomalacia in adults and may result from insufficient exposure to sunlight, insufficient vitamin D in the diet, or both. Severe rickets is characterized by continued collagen-matrix formation and incomplete mineralization, leading to softening of bone. Chronic renal failure can lead to rickets because of decreased formation of 1,25-cholecalciferol by the renal tubules. This is called renal rickets (renal osteodystrophy). Symptoms of rickets can be dramatically improved by consumption of a diet rich in vitamin D or exposure to ultraviolet light. Fish oils, such as those of cod and halibut, are very rich dietary sources of vitamin D. Adequate amounts can also be derived from liver and egg yolk. For adults, the RDA of vitamin D is 200 international units (IU) or 5 mg of cholecalciferol.

Vitamin D toxicity

Toxicity occurs when high daily doses of vitamin D are consumed for several weeks or months. Symptoms include loss of appetite, nausea, thirst, and stupor. Deposition of calcium can occur in many organs, particularly the arteries and kidneys, due to the hypercalcemia caused by increased calcium absorption and resorption of bone.

Figure 10.4

Synthesis of vitamins D_2 and D_3 and biotransformation of vitamin D_3.

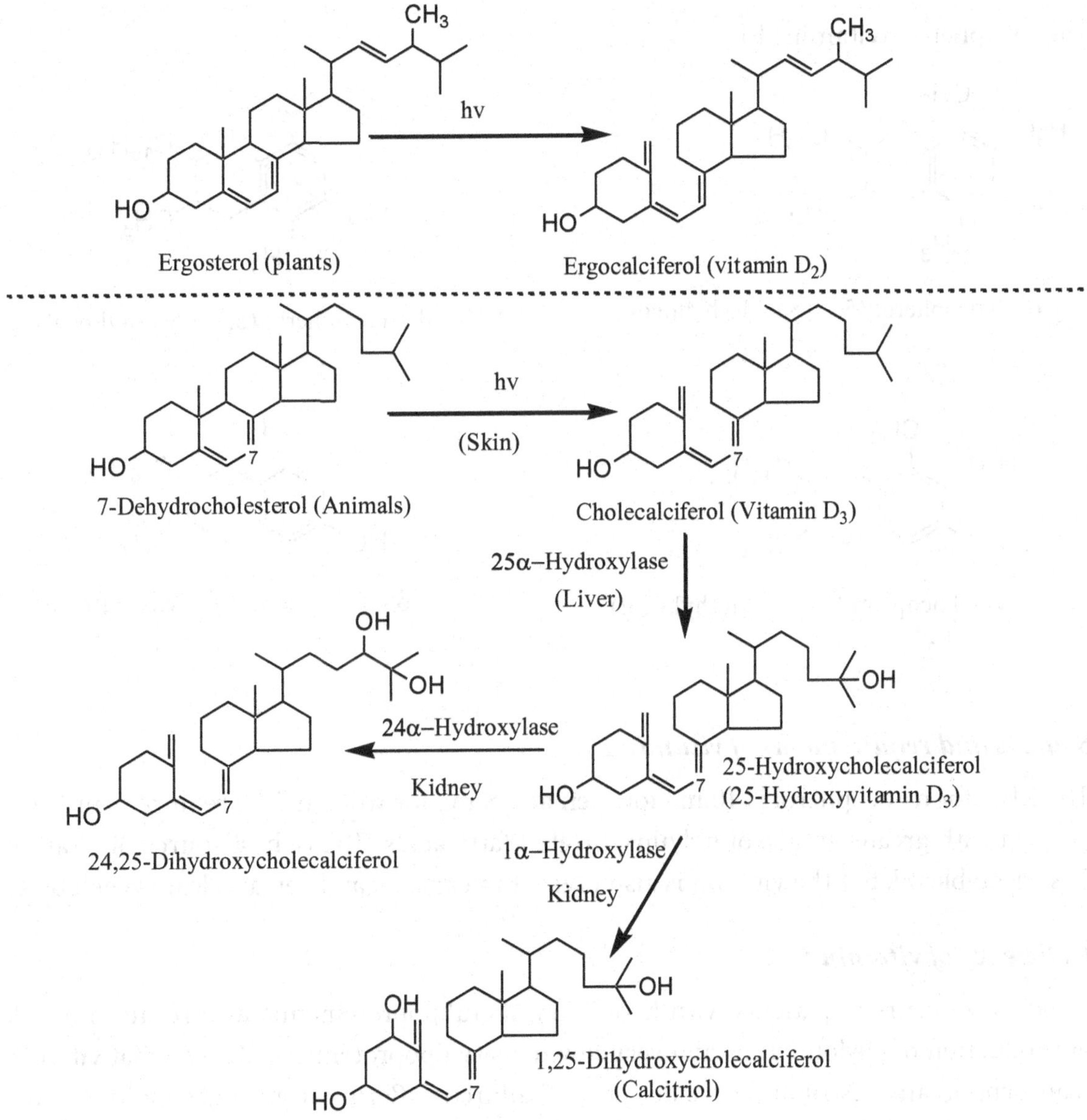

CH3
HO
Ergosterol (plants)
hv
CH3
HO
Ergocalciferol (vitamin D2)
HO
7
7-Dehydrocholesterol (Animals)
hv
(Skin)
HO
7
Cholecalciferol (Vitamin D3)
25α–Hydroxylase
(Liver)
OH
OH
HO
7
24,25-Dihydroxycholecalciferol
24α–Hydroxylase
Kidney
OH
HO
7
25-Hydroxycholecalciferol
(25-Hydroxyvitamin D3)
1α–Hydroxylase
Kidney
OH
OH
HO
7
1,25-Dihydroxycholecalciferol
(Calcitriol)

VITAMIN E (TOCOPHEROL)

Vitamin E refers to several naturally occurring tocopherols, of which α-tocopherol is the most active and most widely distributed. Figure 10.5 shows the structure of some of these compounds. All tocopherols are isoprenoid-substituted 6-hydroxy chromanes or tocols and differ only in the number and position of the methyl groups on the ring. The primary function of vitamin E is as an antioxidant that prevents peroxidation of polyunsaturated fatty acids.

Figure 10.5

The tocopherols (vitamins E).

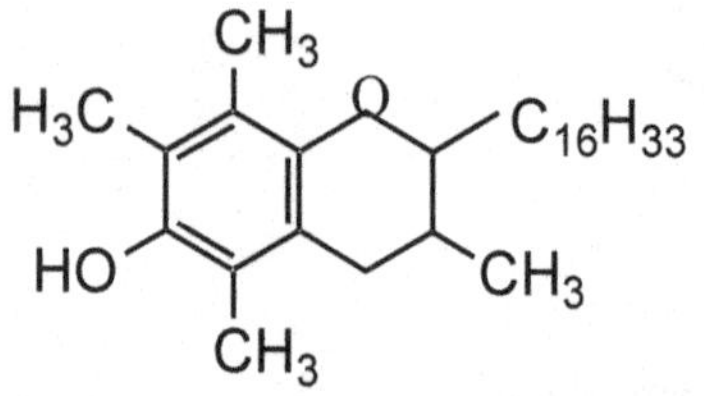

α - Tocopherol (5, 7, 8 - Methyltocol)

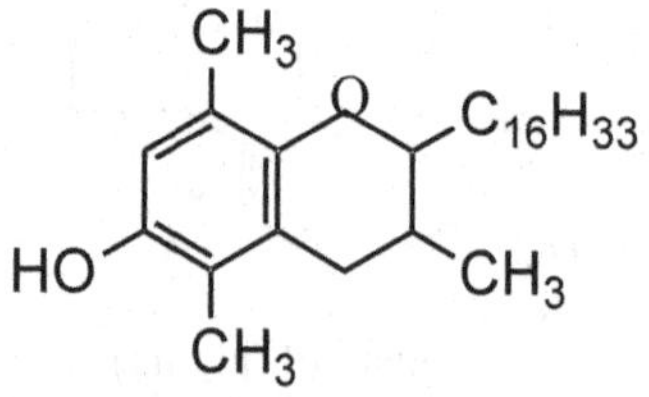

β - Tocopherol (5, 8 - Methyltocol)

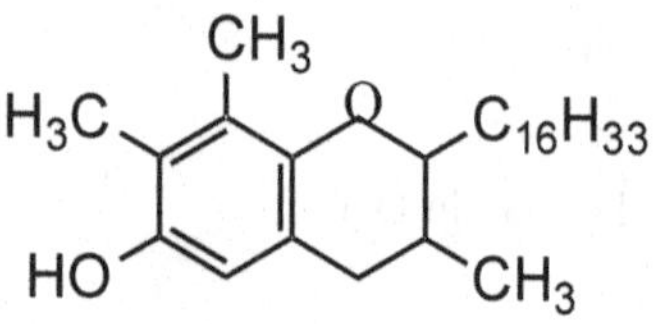

γ - Tocopherol (7, 8 - Methyltocol)

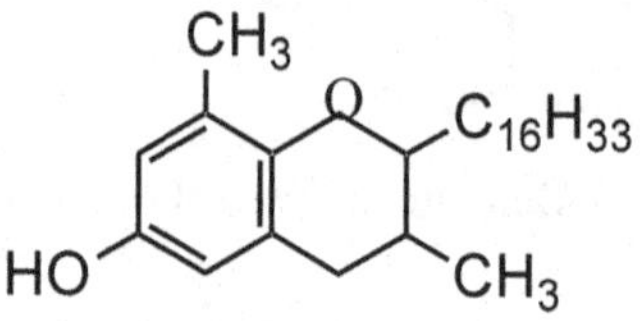

δ - Tocopherol (8 - Methyltocol)

Sources and requirements of vitamin E

The RDA for α-tocopherol is 10 mg for men and 8 mg for women. The requirement is increased with greater intake of polyunsaturated fatty acids. The richest source of vitamin E is vegetable oil, but the vitamin is also present in eggs, meat, liver, and leafy vegetables.

Deficiency of vitamin E

Deficiency occurs in patients with hereditary abetalipoproteinemia as a result of a lack of production of chylomicrons and very-low-density lipoproteins (VLDL) so that vitamin E absorption and distribution are defective. Children suffering from this condition show signs of severe progressive neuropathy and retinopathy. Most patients develop ataxia by the end of the second decade of life. Treatment with high doses of vitamin E can prevent or delay some of these signs. Vitamin E therapy is recommended for all patients with chronic lipid malabsorption syndromes.

In humans, severe vitamin E deficiency leads to neuromuscular abnormalities characterized by spinocerebellar ataxia and myopathy. Similarly, vitamin E deficiency anemia occurs, largely in premature infants, as a result of free-radical damage. Diminished erythrocyte life span and increased susceptibility to peroxide-induced hemolysis are apparent not only in severe deficiency but also in marginal deficiency in hypercholesterolemic subjects. Patients with familial-isolated vitamin E deficiency, an inborn genetic defect in the gene for the α-tocopherol transport protein, have reduced plasma vitamin E levels and characteristic neurological disorders such as cerebellar ataxia, dysarthria, absence of deep tendon reflexes, vibratory and proprioceptive sensory loss, and positive Babinski sign.

Functions of vitamin E

Antioxidant functions

Vitamin E is an important natural antioxidant that protects against peroxidation (auto-oxidation) of polyunsaturated fatty acids (PUFA) in cell membranes of body tissues. Peroxidation occurs when PUFA are exposed to oxygen or free-radical toxicity. Tissue damage may also result from effects on proteins or nucleic acids, which may produce atherosclerosis, inflammatory disease, or cancer. The peroxidation is usually initiated by free-radical formation and is catalyzed by traces of Cu, Fe or hemoglobin, ionizing radiation, or ultraviolet light. Vitamin E reverses the initiation by providing protons and terminates the reactions of peroxidation. Vitamin C is also a radical scavenger and inhibits lipid peroxidation, as does glutathione (GSH).

Figure 10.6 shows lipid peroxidation and its termination by vitamin E. The process is a chain reaction providing a continuous supply of free radicals that initiate further peroxidation. The peroxidation process causes tissue damage. Lipid peroxidation can be classified as enzymatic, non-enzymatic nonradical, and non-enzymatic free-radical mediated. The non-enzymatic free-radical mechanism occurs in three phases: initiation, propagation, and termination.

The process is initiated by the abstraction of H radical from polyunsaturated fatty acid (PUFA) chain to give a free PUFA radical (RO$^\bullet$) in the presence of an initiator. The PUFA radical can react with oxygen to give a peroxy radical of PUFA (ROO$^\bullet$), which in turn reacts with another PUFA to yield a PUFA radical and PUFA hydroperoxide (ROOH). The hydroperoxides produced are strong oxidizing agents – for example, they can oxidize the SH groups of proteins. They are usually disintegrated into fragments including short-chain aldehydes, dialdehydes, and ketones, some of which are considered to be toxic. DNA can also be attacked by the peroxidation process.

Non-antioxidant functions

Cellular signalling

Vitamin E (α-tocopherol) inhibits smooth-muscle-cell proliferation, decreases protein kinase C activity, increases phosphoprotein phosphatase 2A activity, and controls expression of the α-tropomyosin gene. It triggers several events in the host cells, such as the arrest of smooth-muscle proliferation and inhibition of platelet aggregation.

This compound also enhances the release of prostacyclin (PGI_2), a potent vasodilator and inhibitor of platelet aggregation, by enhancing the activity of the rate-limiting enzymes (phospholipase A_2 and the cyclooxygenase) of the arachidonic acid pathway.

Infertility

Vitamin E enhances sperm motility when administered with selenium. The two compounds work synergistically to protect cells from free-radical damage from oxidative stress. However, vitamin E does not replace selenium in prevention of the structural and functional alterations of spermatozoa. Vitamin E supplementation in some infertile women has beneficial effects in improving the endometrial thickness during controlled ovarian stimulation and intrauterine insemination cycles.

Figure 10.6

Lipid peroxidation and its termination by α-tocopherol. The process is started by an initiator with abstraction of H radical from a polyunsaturated fatty acid (PUFA) chain to give a free PUFA radical (RO˙). Vitamin E provides H and terminates the lipid peroxidation.

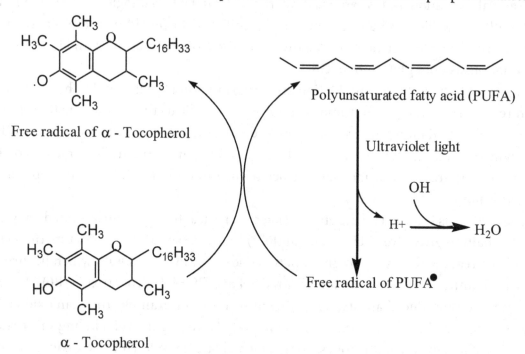

VITAMIN K

Vitamin K is a series of compounds that all have 2-methyl-naphthoquinone nucleus (menadione) but that differ in the isoprenoid side chains at position 3 (figure 10.7). These compounds include phylloquinone (K_1), found in plants; menadione (K_3), a synthetic derivative of vitamin K; and menaquinone (K_2), formed from K_3 by the intestinal flora. In the liver, vitamin K is reduced to vitamin K hydroquinone (KH) by vitamin K reductase, which is sensitive to warfarin. Vitamin K is essential in the post-translational modification of several blood-clotting factors. Absorption of vitamin K requires normal fat absorption. Large doses of K_3 over long periods can cause hemolytic anemia and jaundice in infants.

Dietary sources and requirements of vitamin K

The primary dietary source is phylloquinone derived from green, leafy vegetables and certain vegetable oils (e.g. soybean, rapeseed, and olive oils). The menaquinone form of vitamin K is found in moderate concentrations in only a few types of food, such as cheese and liver. Between 70 and 140 µg of vitamin K per day is an adequate dietary intake.

Biological functions of vitamin K

Vitamin K is essential for the formation of protein C; protein S; blood-clotting factors II, VII, IX, and X; and several proteins involved in bone formation, all of which are synthesized as precursor forms. Formation of biologically active forms of these proteins involves post-translational modification of glutamic acid residues to γ-carboxyglutamate by a specific vitamin K-dependent carboxylase that requires O_2 and CO_2. This process is known as the vitamin K epoxide cycle (figure 10.8).

The 2,3-epoxide product of the carboxylation reaction in the liver is converted into the quinone form by 2,3-epoxide reductase, which is inhibited by warfarin and dicumarol. The quinone is subsequently reduced to the hydroquinone, which is the active form of vitamin K. Vitamin K is used as an antidote to poisoning by dicumarol-type drugs. Dietary vitamin K may enter the cycle via NADPH-dependent vitamin K reductase, which is not affected by inhibitory action of warfarin.

Vitamin K deficiency

Deficiency of vitamin K in adults is unusual because of the wide distribution of the vitamin in the diet and its production by the intestinal flora. However, deficiency can be caused by fat malabsorption or sterilization of the large intestine by orally administered antibiotics. Deficiency in newborn infants leads to hemorrhagic disease. Newborns may develop deficiency because their sterile intestine cannot synthesize vitamin K, because they have a low reserve in their livers since the placenta is poorly permeable to vitamin K, and because human milk contains inadequate amounts compared to recommended formulas for infants.

Figure 10.7

Vitamin K structure.

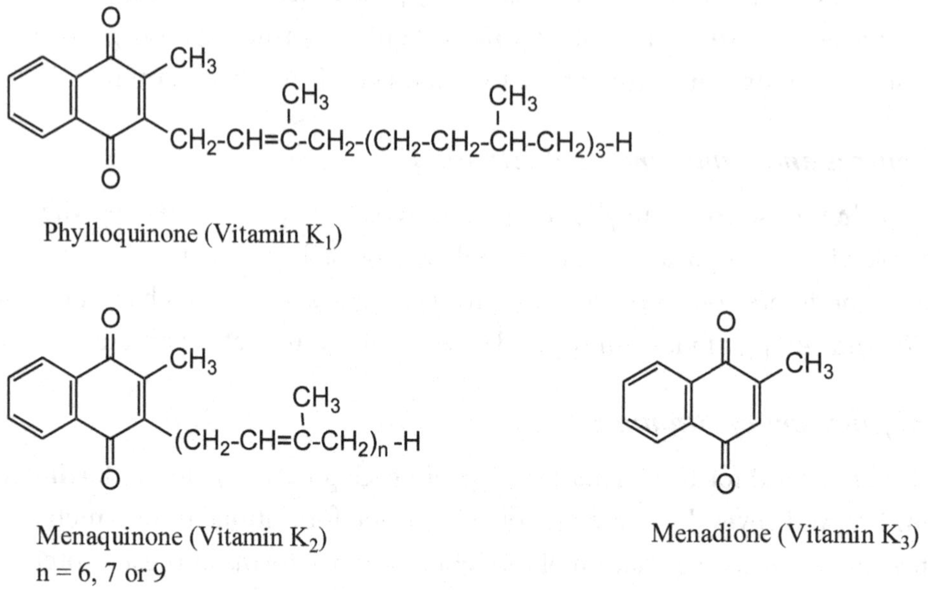

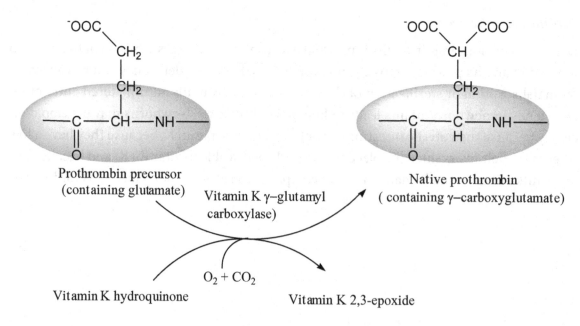

Phylloquinone (Vitamin K_1)

Menaquinone (Vitamin K_2)
n = 6, 7 or 9

Menadione (Vitamin K_3)

Figure 10.8

The vitamin K-epoxide cycle. Dietary vitamin K is converted to hydroquinone (KH) form by vitamin K reductase in the liver. KH is the substrate for vitamin K–dependent carboxylase and vitamin K epoxidase. The γ-carboxylation of glutamic acid residues on the protein substrate is coupled to vitamin K 2,3-epoxide formation in the presence of CO_2 and O_2. Vitamin K quinone is generated from vitamin K 2,3-epoxide by vitamin K epoxide reductase.

Prothrombin precursor
(containing glutamate)

Native prothrombin
(containing γ–carboxyglutamate)

Vitamin K γ–glutamyl
carboxylase)

$O_2 + CO_2$

Vitamin K hydroquinone

Vitamin K 2,3-epoxide

CONCLUSION

- Fat-soluble vitamins A, D, E, and K are essential dietary constituents synthesized from isoprene units in their biological sources.
- They are absorbed and transported like other lipids derived from the diet. Vitamin A is required for vision, vitamin D for calcium and phosphorus metabolism, vitamin E as an antioxidant, and vitamin K for synthesis of prothrombin and several other proteins of blood coagulation.
- Inadequate dietary intake of vitamin A results in night blindness and xerophthalmia, of vitamin D results in rickets in young children and osteomalacia in adults, of vitamin E results in neuromuscular and neurological problems and anemia, and of vitamin K results in bleeding and hemorrhage. Excessive intake of vitamin A or D can result in toxicity.

SELECTED READINGS

Admas, J. S., Clemens, T. L, Parrish, J. A., et al. (1982). Vitamin D synthesis and metabolism after ultraviolet irradiation of normal and vitamin D–deficient subjects. *New England Journal of Medicine* 306:722–725.

Brigelius-Flohe, R., and Traber, M. G. (1999). Vitamin E: Function and metabolism. *The FASEB Journal* 13 (10): 1145–1155.

Davie, E. W. (1995). Biochemical and molecular aspects of the coagulation cascade. *Thrombosis and Haemostasis* 74:1–6.

DeLuca, H. F., and Schones, H. K. (1983). Vitamin D: Recent advances. *Annual Review of Biochemistry* 54: 411–439.

Furie, B., and Furie, B. C. (1990). Molecular basis of vitamin K–dependent γ-carboxylation. *Blood* 75: 1753–1762.

Goodman, D. S. (1984). Vitamin A and retinoids in health and disease. *New England Journal of Medicine* 310: 1023–1031.

Jones, G., Strugnell, S. A., and Deluca, H. F. (1998). Current understanding of the molecular actions of vitamin D. *Physiological Reviews* 78: 1193– 1231.

Sokol, R. J. (1988). Vitamin E deficiency and neurologic disease. *Annual Review ofNutrition* 8: 351–373.

Shearer, M. L., Bach, A., and Kohlmeier, M. (1996). Chemistry, nutritional sources, tissue distribution, and metabolism of vitamin K, with special reference to bone health. *Journal of Nutrtion* 126: 1181S–1186S.

Traber, M. G., and Atkinson, J. (2007). Vitamin E: Antioxidant and nothing more. *Free Radical Biology & Medicine* 43: 4–15.

von Linting, J. (2012). Metabolism of carotenoids and retinoids related to vision. *Journal of Biological Chemistry* 287 (3): 1627–1634.

ABBREVIATIONS

ABC-1, ATP–binding cassette transporter 1

ACAT, acylcholesterol acyltransferase

ACC1, acetyl-CoA carboxylase 1

ACC2, acetyl-CoA carboxylase 2

ACOX1, acyl-CoA carboxylase 1

ACOX2, acyl-CoA carboxylase 2

ACOX3, acyl-CoA carboxylase 3

ACP, acyl-carrier protein

ADP, adenosine diphosphate

AMP, adenosine monophosphate

Apo A, apolipoprotein A

Apo B-48, apolipoprotein B-48

Apo C, apolipoprotein C

Apo D, apolipoprotein D

Apo E, apolipoprotein E

Apo D, apolipoprotein D

Apo I, apolipoprotein I

Apo J, apolipoprotein J

Apo L, apolipoprotein L

AT, acetyl transacylase

ATGL, adipose triglyceride lipase

ACTH, adrenocorticotropic hormone

ATP, adenosine triphosphate

ATPIII, adult treatment panel III

C, cholesterol

cAMP, cyclic adenosine monophosphate

CACT, carnitine-acyl-carnitine translocase

CCK, cholecystokinin

CD-36, cluster differentiation protein-36

CDP, cytidine diphosphate

CDP-DAG, cytidine diphosphate diacylglycerol

CDP-ethanolamine, cytidine diphosphoethanolamine

CE, cholesterol ester

Cer, ceramide

CERT, ceramide transfer protein

CETP, cholesterol ester transfer protein

CHD, coronary heart disease

CHD ChREBP, carbohydrate-responsive element-binding protein

DOC, 11-deoxycorticosterone

CMP, cytidine monophosphate

CM, chylomicron

N-CM, nascent chylomicron

CM-R; chylomicrons remnant

CPT-1, carnitine-palmitoyltransferase I

CPT-2, carnitine-palmitoyltransferase II

CTP, cytidine triphosphate

CVD, cardiovascular disease

CYP4A, cytochrome P4504A

DAG, diacylglycerol

DH, dehydratase

DHAP, dihydroxyacetone phosphate

DHEA, dehydroepiandrosterone

DNA, deoxyribonucleic acid

EFAs, essential fatty acids

EGF, epidermal growth factor

ER, endoplasmic reticulum

ER, enoyl reductase

FA, fatty acyl

FAD, flavin adenine dinucleotide

$FADH_2$, flavin adenine dinucleotide, reduced

FATP, fatty acid transport protein

FAS, fatty acid synthase

FFA, free fatty acid

Gal, galactose

Gal-Cer, galactosylceramide

GalNAc, N-acetylgalactosamine

Glc, glucose

Glc-Cer, glucosylceramide

Glc-CerS, glucosylceramide synthase

GlcNAc, N-acetylglucosamine

GL, glycerolipids

GLC, gas liquid chromatography

GM1, ganglioside GM1

GM2, ganglioside GM2

GM3, ganglioside GM3

GMP, guanosine monophosphate

cGMP, cyclic guanosine monophosphate

GP, glycerophospholipids

GTP, guanosine triphosphate

HDL, high-density lipoprotein

HDL-C, high-density lipoprotein cholesterol

HMG-CoA, 3-hydroxy-3-methylglutaryl-CoA

5-HPETE, 5-hydroxy-6,8,11,14-eicosatetraenoic acid

HRE, hormone-responsive element

HSL, hormone-sensitive lipase

ILCNC, International Lipid Classification and Nomenclature Committee

KH, vitamin K hydroquinone

KR, β-ketoacyl reductase

KS, β-ketoacyl synthase

IDL, intermediate density lipoprotein

IP3, inositol-1,4,5-trisphosphate

LCAT, lecithin: cholesterol acyltransferase

LDL, low-density lipoprotein

LDL-C, low-density lipoprotein cholesterol

LDL-R, low-density lipoprotein receptor

LH, luteinizing hormone

LP(a), lipoprotein (a)

LPX, lipoprotein-X

LT, leukotrienes

LTB_4, leukotriene B_4

LTC_4, leukotriene C_4

LTD_4, leukotriene D_4

LTE_4, Leukotriene E_4

MAG, monoacylglycerol

1-MAG, 1-monoacylglycerol

2-MAG, 2-monoacylglycerol

MCPA, methylenecyclopropyl acetic acid

MGL, monoacylglycerol lipase

mRNA, messenger RNA

MT, malonyl transacylase

MTTP, microsomal triglyceride transfer protein

NAD, nicotinamide adenine dinucleotide

NAD^+, nicotinamide adenine dinucleotide, oxidized

$NADH+H^+$, nicotinamide adenine dinucleotide, reduced

NADP, nicotinamide adenine dinucleotide phosphate

$NADP^+$, nicotinamide adenine dinucleotide phosphate, oxidized

NADPH, nicotinamide adenine dinucleotide, reduced

NAFLD, nonalcoholic fatty liver disease

NCEP, National Cholesterol Education Program

NANA, N-acetylneuraminic acid

NPC1L1, Niemann-Pick C1 like 1

NASH, nonalcoholic steatohepatitis

N-VLDL, nascent VLDL

P450scc CYP11A1, P450 cholesterol side-chain cleavage enzyme

PA, phosphatidic acid

PAF, platelet-activating factor

PAPS, 3-phosphoadenosine-5'-phosphosulphate

PC, phosphatidylcholine,

PE, phosphatidylethanolamine

PG, prostaglandin

PGA_2, prostaglandin A_2

PGD_2, prostaglandin D_2

PGE_2, prostaglandin E_2

$PGF_2\alpha$, prostaglandin $F_2\alpha$

PGG_2, prostaglandin G_2

PGH_2, prostaglandin H_2

PGI, prostacyclin I_2

PI, phosphatidylinositol

PKC, protein kinase C

PLA_1, phospholipase A_1

PLA_2, phospholipase A_2

PLC, phospholipase C

PLD, phospholipase D

PKA, protein kinase-A

PLP, pyridoxal-5-phosphate

PS, phosphatidylserine

PUFA, polyunsaturated fatty acid

RDA, recommended dietary allowance

RE, retinol equivalent

S1P and S2P, site-1 and site-2 protease

SAPs, sphingolipid activator proteins

SCAP, SREBP cleavage activating protein

sdLDL, small dense LDL

SHBG, sex-hormone-binding globulin

SL, saccharolipids
SP, sphingolipids
SR-B1, scavenger receptor class B, type 1
SREBP-1, sterol response-elements binding protein-1
SREBP-2, sterol response-elements binding protein-2
ST, sterol lipid
TAG, triacylglycerols
TE, thioesterase
TLC, thin-layer chromatography
TM, transmembrane domain
TPP, thiamine pyrophosphate
TX, thromboxane
TXA_2, thromboxane A_2
TXB_2, thromboxane B_2
VLDL, very-low-density lipoprotein

GLOSSARY

abetalipoproteinemia: rare autosomal recessive disorder characterized by fat malabsorption, complete absence of plasma apolipoprotein (apo) B-containing lipoproteins.

acetic acid: two-carbon monocarboxylic acid; a product of metabolic breakdown of carbohydrates and fatty acids.

acetyl-CoA: activated form of acetate; the precursor of fatty acids, cholesterol, and ketone bodies.

acetyl-CoA carboxylase: multi-enzyme complex that catalyzes the rate-limiting step of lipogenesis – formation of malonyl-CoA from acetyl-CoA.

acidic glycosphingolipid: negatively charged ganglioside at physiologic pH because it contains n-acetylneuraminic acid (NANA).

acyl-carrier protein: essential component of the fatty acid synthetase complex that carries the acyl moieties during lipogenesis.

acyl-coenzyme: activated form of fatty acids; formed by fatty-acid linkage to coenzyme A with a thiol-ester bond.

acylglycerols: esters of fatty acids with glycerol (e.g. mono-, di-, or triacylglycerol).

Addison's disease: condition of glucocorticoid insufficiency; a result of adrenocortical atrophy or destruction.

adipocyte: cell of adipose tissue specialized for storage of fatty acids as triacylglycerols.

adipose tissue: tissue made up mostly of adipocytes; widely distributed, especially under skin and in the abdominal cavity; serves as an energy store and for protection and insulation.

alcoholism: chronic alcohol consumption that is hepatotoxic and leads to triacylglycerol accumulation and liver cirrhosis.

amphiphilic: containing hydrophobic and hydrophilic portions (e.g. fatty acid).

androgen-binding protein: glycoprotein of cells of seminiferous tubules of testes that binds testosterone, 17-ß-estradiol, and dihydrotestosterone.

antioxidant: compound that inhibits the auto-oxidation of polyunsaturated fatty acids (e.g. α-tocopherol).

apolipoprotein: protein component of plasma lipoproteins that solubilizes the hydrophobic lipid components.

arachidonic acid: an ω-6-unsaturated 20-carbon fatty acid that contains four double bonds and becomes a dietary essential when linoleic acid intake is inadequate.

atherosclerosis: deposition of excess plasma cholesterol and other lipids along the arterial walls with narrowing of the arterial diameter.

auto-oxidation: peroxidation of lipids exposed to oxygen; responsible for the deterioration of foods (rancidity) and human tissue damage in vivo.

Babinski reflex: upward movement of the big toe and fanning of other toes when sole of the foot is stroked firmly.

bile acid: acid formed in hepatocytes from cholesterol and conjugated with glycine or taurine; bile salts (sodium glycocholate and sodium taurocholate) are required for digestion and absorption of dietary lipids.

branched-chain fatty acid: straight-chain fatty acid with one or more methyl substituents (e.g. phytanic acid).

butyric acid: saturated 4-carbon monocarboxylic fatty acid.

calcitriol: 1,25-dihydroxycholecalciferol, the active form of vitamin D_3; formed in the kidneys from 25-hydroxycholecalciferol by 1-α-hydroxylase.

cardiolipin: diphosphatidylglycerol; localized primarily in the mitochondrial inner membrane.

carnitine: ß-hydroxy-γ-trimethylammonium butyrate; required for transport of acyl-CoA moieties into the mitochondrial matrix.

carnitine deficiency: impaired energy production from long-chain fatty acids, particularly during periods of fasting or stress; results from a lack of carnitine.

carnitine shuttle: transport process of acyl-CoA moieties into the mitochondrial matrix.

ß-carotene: pro-vitamin A; an isoprenoid compound that occurs widely in vegetables (e.g. carrots) and contains the equivalent of 2 molecules of retinol.

ceramide: n-acyl-sphingosine; an essential component of sphingolipids.

cholelithiasis: formation of cholesterol gallstones.

cholesterol: sterol with 27 carbon atoms; formed in the human body from acetyl-CoA and involved in control of cell-membrane fluidity and the precursor of steroid hormones, bile salts, and vitamin D_3.

choline: n-trimethylethanolamine; an essential nutrient required for formation of phosphatidylcholine.

chromatography: technique for separation and identification of the component of a mixture, such as lipids.

chylomicron (CM): the least dense and largest of the plasma lipoproteins.

corticosteroid-binding globulin (CBG): transcortin; plasma ß-globulin that transports glucocorticoids and progesterone.

congenital adrenal hyperplasia (CAH): condition of hypoadrenocorticism due to a genetic deficiency of an enzyme involved in steroid-hormone synthesis.

Cushing's syndrome: condition of excess glucocorticoid production by the adrenal medulla.

cytidine triphosphate (CTP): energy-rich nucleotide that drives certain reactions in glycerosphingolipid synthesis and a source of cytidine-monophosphate units of DNA and RNA.

dicarboxylic aciduria: excretion in urine of dicarboxylic acids from lack of mitochondrial medium-chain acyl-CoA dehydrogenase.

1,25-dihydroxycholecalciferol: *see* calcitriol.

eicosanoid: derivative of an eicosa-(C20) polyunsaturated fatty acid; may be a prostanoid or a leukotriene.

eicosapentaenoic acid: ω-3-unsaturated 20-carbon fatty acid that contains 5 double bonds.

emulsification: dispersion of a hydrophobic material into an aqueous solution or vice versa by an agent such as a bile salt.

enterohepatic circulation: circulation of bile acids from the liver into the small intestine and back to the liver.

ergocalciferol: active form of vitamin D_2 formed from ergosterol (provitamin D_2) in plants.

ergosterol: provitamin D_2; differs from 7-dehydrocholesterol (provitamin D_3) in that its side chain is unsaturated and contains an extra methyl group.

essential fatty acid: linoleic acid and α-linolenic acid which cannot be synthesized by the body and must be supplied in the diet.

estrogen: steroid hormone (e.g. estrone, estradiol, and estriol) secreted by the ovaries and placenta during pregnancy upon stimulation by follicle-stimulating hormone.

fatty acid: aliphatic monocaboxylic acid component of complex lipids. The carbon chain varies in length, may be saturated or unsaturated, and may carry functional groups containing oxygen, a halogen, nitrogen, or sulphur.

fatty acid synthase complex: a dimer of 2identical multifunctional polypeptides arranged in an antiparallel configuration, each having 7 enzymatic activities and an acyl carrier function.

follicle-stimulating hormone (FSH): protein hormone secreted by the anterior pituitary gland that acts on granulosa cells of the ovaries and sertoli cells of the testes.

galactocerebroside: neutral glycosphingolipid; consists of ceramide and a galactose unit.

gallstones: accretions of cholesterol and/or calcium bilirubinate that form in the gall bladder.

ganglioside: complex glycosphingolipid found primarily in ganglion cells of the central nervous system; consist of a ceramide oligosaccharide and contain one or more molecules of n-acetylneuraminic acid.

glucagon: polypeptide hormone secreted by the α-cells of the pancreas in response to low blood glucose to increase its level.

glucocerebroside: neutral glycosphingolipid; consists of ceramide and a glucose unit.

glucocorticoid: steroid hormone (e.g. cortisol) synthesized from progesterone in the adrenal zona fasciculate; promotes gluconeogenesis and breakdown of fat and protein.

gluconeogenesis: formation of glucose from non-carbohydrate sources such as alanine or glycerol.

glucosylceramide: glycosphingolipid which consists of ceramide and glucose.

glyceroglycan: glycerol linked to one or more monosaccharide units by glycosidic linkage.

glycerol kinase: converts glycerol and ATP into glycerol 3-phosphate and ADP.

glycerol 3-phosphate: phosphoric monoester of glycerol; the precursor of acylglycerols.

glycerophospholipid: glycerolphosphate that contains 1 or 2 fatty acids esterified to the glycerol fraction.

glycolysis: metabolic pathway in which glucose is converted into 2 molecules of pyruvate. The process is coupled to the production of ATP and provides metabolite precursors for other biomolecules.

glycosphingolipid: derivative of a ceramide with one or more monosaccharides.

glycosphingolipid-storage disease: sphingolipidosis; inherited disorder in which a specific glycosphingolipid accumulates in cells and tissues because of a defect in degradation of sphingolipids.

gonadotropin releasing hormone: luteinizing-hormone-releasing hormone; peptide hormone formed in the hypothalamus that regulates release from the anterior pituitary gland of luteinizing hormone and follicle-stimulating hormone.

high-density lipoprotein (HDL): smallest of plasma lipoproteins; consists mostly of protein and is formed in the liver and small intestine.

hormone: regulatory molecule secreted by an endocrine gland to influence target cell functions.

hormone-responsive element (HRE): regulatory sequence of DNA, that binds a nuclear receptor-hormone complex.

hormone-sensitive lipase (HSL): enzyme in adipose tissue that hydrolyzes triacylglycerol into free fatty acids and glycerol for release into the blood.

3-hydroxybutyrate: a ß-hydroxybutyrate formed by reduction of acetoacetate with aceton and acetoacetate, one of the ketone bodies.

hypercholesterolemia: elevated plasma cholesterol level, usually elevated plasma low-density lipoprotein level.

hyperlipoproteinemia: elevated level of one or more plasma-lipoprotein fractions.

hypervitaminosis A: a condition caused by excessive intake of vitamin A characterized by headache, loss of appetite, hepatosplenomegaly, and dermatitis.

hypotonia: condition of low muscle tone.

inhibin B: glycoprotein released by granulosa cells in the ovaries that decreases secretion of follicle-stimulating hormone.

inositol-1,4,5-trisphosphate: compound released from phosphatidyl-4,5-bisphosphate by phospholipase C as a second messenger upon hormonal stimulation.

insulin: protein hormone secreted by the ß-cells of the pancreas in response to high blood glucose.

Jamaican vomiting sickness: illness caused by eating the unripe fruit of the ackee tree, which contains the toxin hypoglycin, which inhibits ß-oxidation of fatty acids.

ketoacidosis: condition caused by elevated blood levels of the ketone bodies acetoacetic acid and ß-hydroxybutyric acid, which lower the blood pH, in uncontrolled type 1 diabetes mellitus.

ketogenesis: synthesis of ketone bodies from acetyl-CoA by hepatocytes when fat and carbohydrate degradation are not appropriately balanced.

ketone body: acetoacetate, ß-hydroxybutyrate, or acetone exported from the liver into the blood. In extrahepatic tissues, the first two are used as sources of energy.

ketonemia: increased level of ketone bodies in the blood.

ketonuria: increased level of ketone bodies in the urine.

ketosis: increased level of ketone bodies in the blood and urine.

lecithin: phosphatidylcholine.

lecithin-cholesterol acyltransferase (LCAT): enzyme responsible for the formation of most plasma cholesterol esters from lecithin and cholesterol.

leukotriene: eicosanoid characterized by the presence of 3 conjugated double bonds and formed by lipoxygenase.

lingual lipase: acid-stable lipase secreted by small glands (Ebner's glands) in the back part of the tongue; releases short-chain fatty acids from position 3 of triacylglycerol of milk.

linoleic acid: essential ω-6 unsaturated fatty acid.

α-linolenic acid: essential ω-3 unsaturated fatty acid .

lipase: one of a group of enzymes that hydrolyze acylglycerol ester bonds.

lipid: organic compound that is water insoluble but that is soluble in nonpolar solvents such as chloroform or ether.

lipid digestion: process by which dietary fat from animal or plant sources is broken down into its components in the alimentary tract.

lipid malabsorption: steatorrhea; excessive loss of dietary lipids in the stool from a defect in pancreatic lipase secretion, bile salt release, or in intestinal mucosal cells.

lipid-storage disease: sphingolipidosis.

lipogenesis: de novo synthesis of palmitate from acetyl-CoA.

lipolysis: hydrolysis of lipids, especially triacylglycerol, into free fatty acids and glycerol.

lipoprotein: complex of a variety of lipids with one or more specific proteins called apolipoproteins.

lipoprotein (a): complex plasma lipoprotein; a variant of low-density lipoprotein that contains an apolipoprotein (a) covalently linked by a disulphide bond to apo B-100.

lipoprotein lipase: enzyme that hydrolyzes triacylglycerol of plasma lipoprotein into free fatty acids and glycerol.

lipoprotein-X: abnormal plasma low-density lipoprotein found in individuals with intra- or extra-hepatic cholestasis, familial lecithin cholesterol acyltransferase deficiency, and newborn infants with immature liver function.

lipotropic factor: substance that prevents or corrects abnormal accumulation of fat in the liver (e.g. choline).

lipoxygenase: iron-containing enzyme that catalyzes oxidation by oxygen of polyunsaturated fatty acids to hydroperoxides in a stereo-specific manner.

lipoxygenase pathway: process of formation of leukotrienes from arachidonic acid.

lovastatin: polyisoprenoid statin drug that inhibits 3-hydroxy-3-methylglutaryl-CoA reductase and lowers elevated plasma cholesterol level.

low-density lipoprotein (LDL): plasma lipoprotein formed through the action of lipoprotein lipase on very-low-density lipoprotein; it is rich in cholesterol, which it transports to peripheral tissues.

luteinizing hormone (LH): protein hormone secreted by the anterior pituitary gland that acts on Leydig cells of the testes.

lysophosphoglyceride: phosphoglyceride that lacks the fatty acid at either carbon 1 or 2.

malonyl-CoA: formed by carboxylation of acetyl-CoA during lipogenesis; inhibits carnitine-palmitoyl-transferase I, thereby inhibiting ß-oxidation of fatty acids.

mevalonate: a 6-carbon intermediate in cholesterol synthesis produced by condensation of 3 acetate units.

mevastatin: statin drug that inhibits 3-hydroxy-3-methylglutaryl-CoA reductase, thereby lowering plasma cholesterol level.

micelle: water-soluble aggregate of amphipathic lipids with their polar groups oriented towards the aqueous medium and their hydrophobic regions towards the centre of the particle.

microsomal triglyceride transfer protein: protein of the endoplasmic reticulum; essential for formation of very-low-density lipoprotein in the intestinal mucosa and liver.

mineralocorticoid: steroid hormone secreted by the adrenal zona glomerulosa such as aldosterone, which stimulates reabsorption of sodium and excretion of potassium and H+ by the renal tubules.

monoacylglycerol: ester of glycerol with one fatty acid unit.

neutral glycosphingolipid: ceramide monosaccharide that contains either a galactose (galactocerebroside) or glucose (glucocerebroside) unit.

night blindness: inability to see at night or in dim light because of vitamin A deficiency.

nucleotide: molecule that consist of a purine or pyrimidine, a ribose or 2-deoxyribose, and a phosphate group.

obesity: condition in which there is an excessive amount of body fat.

oleic acid: monounsaturated, 18-carbon fatty acid.

opsin: light-sensitive membrane-bound G-protein found in retinal cells.

osteomalacia: softening and deformation of bone in adults as a result of vitamin D deficiency.

α-oxidation: oxidation of a ß-methyl fatty acids at the α carbon as a first step in its metabolism, as in the case of phytanic acid.

ß-oxidation: successive oxidation of fatty acids at the ß-carbon with removal of 2-carbon fragments as acetyl-CoA.

ω-oxidation: oxidation of the methyl terminal (ω-carbon) in microsomes of fatty acids between chain-length C10 and C26 fatty acids in microsomes to form a dicarboxylic fatty acid that can then undergo ß-oxidation from either end.

oxidation of fatty acids: degradation process by which fatty acids are converted into acetyl-CoA in mitochondria, mainly by ß-oxidation, which is coupled to the generation of energy.

palmitic acid: saturated 16-carbon fatty acid.

palmitoyl-CoA: activated form of palmitate.

pancreatic lipase: hydrolytic enzyme secreted by pancreas that catalyzes the removal of fatty acids at carbons 1 and 3 of triacyglycerol to produce 2-monoacylglycerol.

peroxidation: oxidative degradation of cell-membrane polyunsaturated fatty acids exposed to either oxygen or free radicals; could cause tissue damage.

peroxisomal oxidation: the ß-oxidation of very-long-chain fatty acids in peroxisomes in which flavoprotein oxidase passes electrons directly to O_2 to produce H_2O_2.

phosphatidic acid (PA): diacylglycerolphosphate.

phosphatidylcholine (PC): lecithin; glycerophospholipid with choline as its polar group.

phosphatidylethanolamine (PE): cephalin; glycerophospholipid with ethanolamine as its polar group.

phosphatidylinositol (PI): glycerophospholipid with inositol as its polar group.

phosphatidylserine (PS): glycerophospholipid with serine as its polar group.

phospholipase: enzyme that hydrolyzes an ester bond in aglycerophospholipid; such enzymes are classified according to the bond cleaved.

phospholipase A_1: enzyme that removes fatty acid from the sn-1 position of a glycerophospholipid.

phospholipase A_2: enzyme that removes fatty acid from the sn-2 position of a glycerophospholipid.

phospholipase C: enzyme that removes the phosphorylated nitrogen base from a glycerophospholipid.

phospholipase D: enzyme that removes the nitrogenous base from a glycerophospholipid.

phospholipid: lipid that contains esterified phosphoric acid (e.g. glycerophospholipid and sphingophospholipid).

photolysis: chemical decomposition of a compound by a photon of light or radiant energy.

phylloquinone: vitamin K_1, present in green, leafy vegetables and certain vegetable oils.

phytanic acid: 3,7,11,15-tetramethylhexadecanoic acid formed from the phytol component of chlorophyll.

plasmalogen: glycerophospholipid with a fatty acid attached by ether rather than by an ester linkage at carbon 1 of the glycerol moiety (e.g. phosphatidal ethanolamine).

platelet-activating factor: ether lipid 1-alkyl-2-acetylglycerophosphocholine; a potent platelet-aggregating agent.

polyketide: poly-beta-ketone; a compound such as lovastatin that contains more than 2 carbonyl groups separated by a carbon atom and that is produced from acetyl-CoA by a variety of organisms.

polyisoprenoid: a compound such as cholesterol made up of more than one isoprene unit.

polyunsaturated fatty acid: long-chain fatty acid that contains more than one double bond.

propionic acid: 3-carbon monocarboxylic fatty acid.

prostacyclin-I$_2$: a prostaglandin formed by vascular endothelium that inhibits platelet aggregation and causes vasodilation.

prostaglandin: prostanoid that acts as a local hormone.

prostanoid: derivative of an eicosa-(C20) polyunsaturated fatty acid that functions as a regulatory molecule (e. g. prostaglandin).

prothrombin: coagulation factor required for formation of a blood clot.

recommend dietary allowance (RDA): amount of a nutrient to be ingested daily in the diet to maintain normal health.

Refsum's disease: severe neurological condition with accumulation of phytanic acid in tissues and body fluids due to a genetic deficiency of phytanoyl-CoA hydroxylase .

retinoid: isprenoid that contains four isoprene units, such as natural or synthetic forms of vitamin A.

retinol: vitamin A; a polyisoprenoid that contains a cyclohexenyl ring.

retinol equivalent (RE): biological activity of retinol, with 1 RE equal to 1 µg of all-*trans* retinol.

rhodopsin: photosensitive pigment in rod cells of the retina that consists of 11-*cis*-retinal and the protein opsin.

rickets: condition of softening and deformation of bones in children due to deficiency of vitamin D.

saccharolipid: any lipid that has a carbohydrate component; also used specifically for lipids that consist of fatty acids esterified to one or more carbohydrate components.

saturated fatty acid: straight-chain fatty acid with no double bonds (and a hydrophobic tail).

Schiff base: formed by condensation of an amine group with a carbonyl group of an aldehyde or ketone.

secretin: polypeptide hormone secreted by the duodenum in response to dietary fat that stimulates secretion of pancreatic juices.

sex-hormone-binding globulin: plasma glycoprotein secreted by the liver that binds more tightly to androgens and estrogen than to albumin.

slow-reacting substance of anaphylaxis: mixture of leukotrienes formed in anaphylaxis that causes contraction of smooth muscle.

sphingolipid: a lipid that consists of sphingosine, to which are attached a long-chain fatty acid in amide linkage and a polar group.

sphingolipidosis: inherited disorder of sphingophospholipid degradation with accumulation of a specific lipid in cells and tissues.

sphingomyelin: important sphingolipid in mammals in which the polar group is phosphorylcholine.

sphingosine: D-erythro-sphingosine or 4-sphingenine; a long-chain aminoalcohol formed from a palmitic acid and a serine.

steroid hormone: hormone derived from cholesterol in the testes, ovaries, or adrenal cortex that exerts its effects in the nucleus of its target cells.

sulphatide: the glycolipid sulphated galactosyl-ceramide.

thromboxane: oxygenated eicosanoid formed by blood platelets that stimulates platelet aggregation and causes vasoconstriction.

tocopherol: polyisoprenoid; form of vitamin E.

transcortin: corticosteroid-binding globulin.

transducin: heterotrimeric G-protein important in phototransduction in the retina.

triacylglycerol (TAG): triglycerides; esters of three fatty acids with glycerol.

triglyceride: triacylglycerols.

unsaturated fatty acid: fatty acid that contains one or more double bonds.

very-low-density lipoprotein (VLDL): plasma lipoprotein formed and secreted by the liver and modified in the circulation into intermediate-density lipoprotein and low-density lipoprotein particles.

xerophthalmia: condition of dryness of the conjunctiva and cornea caused by vitamin A deficiency.

INDEX

Note: page numbers followed by *f* denote figures, while page numbers followed by *t* denote tables.